THE FOUL
AND THE FRAGRANT

The Foul

ODOR AND THE

Cambridge, Massachusetts   1986

and the Fragrant

FRENCH SOCIAL IMAGINATION

Alain Corbin

HARVARD UNIVERSITY PRESS

Typeset in Linotron Garamond No. 3 and
designed by Copenhaver Cumpston

Library of Congress Cataloging-in-Publication Data

Corbin, Alain.
    The foul and the fragrant.

    Translation of: Le miasme et la jonquille.
    Includes bibliographical references and index.
    1. Odors—Social aspects.   2. Smell—Social aspects.
3. France—Social conditions—18th century.   4. France—
Social conditions—19th century.   I. Title.
GT2847.C6713   1986   394   86-7684
ISBN 0-674-31175-2 (alk. paper)

Foreword

TODAY'S HISTORY comes deodorized. Thanks to experts in art, architecture, and artifacts, our eyes have been opened to what the past looked like; and all who have immersed themselves in diaries, novels, and letters will have their ears attuned to the distant sounds of civilized life. But how many historians have given us the smell of previous societies? Researchers have been all too silent, repelled, it seems, by modern hygienic sensibilities even from contemplating the stench of the past. Smell—both as an emanation of material culture and as part of the empire of the senses—though fundamental to experience has been neglected by scholars.

Alain Corbin has now made a memorable contribution to putting smell on the historical map, in a book whose erudition and originality have already ensured it a dazzling impact in France. Integrating an impressive range of specialist disciplines—not least, the histories of science and of medicine, urban studies, public health, psychohistory, and literary criticism—he conjures up the dominion exercised in past time over people's lives—and deaths—by the combined forces of smells, from the seductress's civet to the ubiquitous excremental odors of city cesspools.

Corbin succeeds in capturing all such assaults on the nose and the viscera as they struck the French in the eighteenth and nineteenth centuries. His ambitions, however, are far greater: exploring the deeper repercussions of smell for science, society, and literature. In the "world we have lost" before today's hygienic regimes, stenches filled the nose; but they also filled the mind. Smell featured crucially in leading theories of life, disease, and the atmosphere and in technologies of health from the Enlightenment to the mid-nineteenth-century heyday of the sanitarians. Pre-Pasteurian orthodoxy held that sickness arose from pestilential miasmas given off by the environment, by towns, and by their fetid populations. Stench was, in fact, disease. And as experts increasingly sniffed out the sources of stench among the "great unwashed," sanitary reformers and social engineers joined forces in campaigns against filth in all its modes—physical, moral, and verbal. In other words, argues Corbin, public health must be seen as more than a milestone on the road of progress; it had its wider politics as one of the Foucaultian disciplines of social control.

So, to borrow Mary Douglas's formula, the "purity" of disinfection could wage successful battle against the "danger" of pollution. Yet, paradoxically—as Corbin brilliantly reveals—smell remained potent and alluring. In the psycho-semiology of olfaction, body odors spelled protection, strength, and sexuality. Not least, one's smell was a unique "fingerprint" of identity. Hence, although the literary imagination from Balzac to Zola certainly picked on smell to signal disgust toward the masses, through Romanticism it also evoked the mysteries of love, reverie, and memory. In contrast to the fecal stink of the herd, argues Corbin, there was the unique fragrance of the loved one: pure, natural. The outcomes were material as well as literary: the art of the *parfumier*, in a Coty, Worth, or Guerlain, ceased to be one of masking and became the secret means of revealing inner identity. And if, for Flaubert and for Huysmans above all, the power of smell was delicious, decadent, and ultimately destructive, dredging up from the depths a welter of animal associations, isn't that merely to say that the artist had already anticipated Freud's discovery of smell as the meeting point of desire and shame within the paradox of civilization?

Corbin traces the intellectual affinities of smell from science, via salubrity, to sensibility, uncovering connections of keen interest to specialists in history, literature, and medicine no less than to every reader fascinated by the cultural anthropology of the senses. Combining that commitment to "total history" pioneered by the Annales school, with a nose for pungent detail, he has called up a lost world and excitingly enriched our perceptions of the past.

ROY PORTER
Wellcome Institute for the
History of Medicine, London

Contents

Miriam L. Kochan provided a first draft translation, and Dr. Roy Porter and Christopher Prendergast contributed their knowledge and skills to achieve the final version. The publishers gratefully acknowledge their help.

THE FOUL
AND THE FRAGRANT

Introduction

THE IDEA OF writing a book about the perception of odors came to me as I was reading the memoirs of Jean-Noël Hallé, a member of the Société Royale de Médecine under the ancien régime and the first incumbent of the chair of public hygiene established in Paris in 1794.

Let us follow him through three episodes of his tireless struggle, beginning with February 14, 1790. More than six months have passed since the storming of the Bastille. The situation has calmed down. Mild temperatures announce the end of winter. On this day the thermometer rises to four degrees Réaumur; there is a southeasterly breeze; near the Pont de la Tournelle the Seine reaches a level of five feet. Accompanied by his friend Boncerf, Jean-Noël Hallé has gone out early in the morning to explore the odors on the riverbanks, or rather, to sniff them out. Both scientists have been entrusted with this task by the Société Royale de Médecine. Starting at the Pont Neuf, they stride along the right bank up to La Rapée. They cross the river almost opposite the sewer of the Salpêtrière in order to return on the left bank to their departure point. The meticulous record of their walk of more than ten kilometers provides an accurate picture of the variety of odors. There is no reference anywhere in the text to anything visual.

Hallé's report makes rather unsatisfactory reading for amateurs of the picturesque; similarly, there is no mention of the chatter of washerwomen or of the noisy bustle of dockers on either side of the Seine. Nothing but odors—the discontinuous mapping of a walk that favors rubbish and ignores those parts of the river where the quays or houses are built directly into the water and where no stench can be perceived.

This olfactory survey is not without danger. One incident shows how necessary it is to guard against excessive boldness. At the mouth of the dreadful Gobelin tributary Hallé's companion, *facing the breeze,* walks along the edge of the water and wades through the black mud.

> Monsieur Boncerf, who at this point had turned more directly into the southeasterly breeze and had descended to the riverbank, was overcome by a biting, alkaline, stinging, and stinking odor. It affected his respiratory system so badly that his throat began to hurt within half an hour and his tongue became noticeably swollen. Affected by these poisonous vapors, he warned me to return to the road straightaway; because I had remained at the easternmost point of the bank that had been infested by these sediments, and hence with my back to the wind, I myself did not experience anything unpleasant.[1]

But this was no more than a minor skirmish. The long battle with stench provided incomparably more dramatic episodes. Consider another incident eight years earlier. On March 23, 1782, the most famous experts in hygiene and chemistry gather in front of the Hôtel de la Grenade in the rue de la Parcheminerie. The cesspool of the building is to be cleaned out. The fatal character of its effluvia is well known. Moreover, the landlady is certain that the medical students have buried beneath the feces arms, legs, and other parts of the human body by the bucketful. The extent of the danger is without precedent. The Académie Royale des Sciences has dispatched the academicians Lavoisier, Le Roy, and Fougeroux to make an on-the-spot inspection. The chemists Macquer and Fourcroy and the duc de La Rochefoucauld, the abbé Tessier, and Jean-Noël Hallé have come by order of the Société Royale de Medécine. They all are supposed to test the effectiveness of a new antimephitic substance; its inventor, Sieur Janin, has daringly asserted that it will destroy foul smells and quell miasmas.

It is a cold day, a mere two degrees Réaumur around noon. The wind is blowing from the north. There has been a heavy snowfall during the morning. In short, meteorological conditions appear favorable. While Janin sprinkles his substance, Jean-Noël Hallé and Abbé Tessier climb up and down ladders in order to measure the varying intensity of the stench. For hours the experiment, which began between eight and nine in the morning, proceeds without incident. Then, around three in the afternoon, comes a dramatic turn of events: one of the cesspool cleaners suffers a fit of asphyxiation and slips off the ladder into the cesspool. He is pulled out with extreme difficulty.

The onlookers lean over the mortally ill cleaner. A young man tries in vain to save him by administering artificial respiration. At this point Monsieur Verville, an expert, intervenes. He is an inspector for a company that makes the ventilators which have been used for some years during the cleaning of cesspools. Hallé graphically describes the fate of the hapless Verville.

> He had scarcely inhaled the air that was coming from the mouth of the mortally ill man when he shouted "I am a dead man!" and fell down unconscious . . . I saw that he was making an extreme effort to regain his breath; he was held by the arms, as he reared up with a loud groan; his chest and his stomach moved up and down alternately in violent convulsive movements. He had lost consciousness; his limbs were cold; his pulse became weaker and weaker . . . Occasionally his mouth even filled with foam, his limbs became stiff, and the sick man appeared to be having a genuine epileptic fit.[2]

Fortunately Monsieur Verville regains consciousness. He has done no more than inhale the breath of a mortally ill person, and he is soon able to go home. But for a long time he suffers from aftereffects. As he explains, "cesspool gases that have been transmitted" are even more terrifying than those that threaten to suffocate the cleaner at the bottom of the cesspool.

Stay with Jean-Noël Hallé a moment longer, this time to accompany him on his general medical practice. The following report is highly detailed, but not a single word should be omitted; it describes the various pathogenic odors that develop in that olfactory inferno, the hospital.

> There is a stench that is similar to the one exuded by clothes, and there is a moldy smell that is less noticeable but nevertheless more

unpleasant because of the general revulsion it arouses. A third, which might be called the odor of decomposition, may be described as a mixture of the acidic, the sickly, and the fetid; it provokes nausea rather than offending the nose. This mixture accompanies decomposition and is the most repellent among all the odors to be encountered in a hospital. Another odor, which makes the nose and eyes burn, results from uncleanliness. It gives the impression that the air contains something like powder, and, if one looks for the source, one is certain to find damp moldy laundry, a pile of rubbish or clothes and bedclothes infested by fermenting miasmas. Each infectious material has its distinctive exhalation. Doctors know the special smell of a septic wound, of a cancerous agent, and the pestilential odor that is spread by caries. But what physicians have learned of this subject from experience can be tested by anyone who compares the various odors in the wards. In the pediatric ward the smell is sour and stinking; in the women's wards it is sweet and putrid; the men's wards, on the other hand, exude a strong odor that merely stinks and hence is not so repulsive. Although there is now a greater emphasis on cleanliness than in earlier times, in the wards occupied by the good poor of Bicêtre there prevails a flat odor that produces an effect of nausea in delicate constitutions.[3]

Hallé's statements and behavior are by no means unusual. A careful reading of contemporary texts reveals a collective hypersensitivity to odors of all sorts. Pleasure at the sight of the landscape of an English garden or the blueprint of an ideal city is paralleled by horror of the city air, which is infested by miasmas.[4] But we moderns risk an anachronistic perspective. From the anguished quest of Jean-Noël Hallé onward, something changes in the way smells are perceived and analyzed. This book traces the course of that change.

What is the meaning of this more refined alertness to smell? What produced the mysterious and alarming strategy of deodorization that causes us to be intolerant of everything that offends our muted olfactory environment? By what stages has this far-reaching, anthropological transformation taken place? What are the social stakes? What kinds of interests are behind this change in our evaluative schemas and symbolic systems concerning the sense of smell?

Lucien Febvre was among the first to recognize the problem; the history of olfactory perception is one of the many avenues of investigation that he has opened up.[5] Since then historians have turned their attention to the senses of sight and smell; in the case of sight, a project stimulated by the discovery of the great panoptic dream (its claim to attention further strengthened by developments in aesthetic theory); in the case of taste, a project sheltered behind the desire to analyze the rituals and forms of sociability of everyday life. In this sphere as well, the sense of smell has suffered from an unremitting process of discrediting since the time when a new offensive against the olfactory intensities of public space was being sketched out.[6]

The time has come to trace the conflict-laden history of the perception of smells and to study the logic of the systems of images from which that history has been generated. But it is also necessary to relate divergent modes of perception to social structures. It would be futile to analyze social tensions and conflicts without accounting for the different kinds of sensibilities that decisively influence them. Abhorrence of smells produces its own form of social power. Foul-smelling rubbish appears to threaten the social order, whereas the reassuring victory of the hygienic and the fragrant promises to buttress its stability.

Initially there seems to be considerable accord between Hallé's behavior and the philosophical positions of his time. The subtle sensitivity with which he approaches the phenomena of sensory perception reflects the influence of sensualism upon scientific method. Sensualist theory, which was based on the thought of Locke, had been developed in basic outline by Maubec as early as 1709.[7] A generation later it had been given greater precision by David Hartley, whose writings were translated into French in 1755. The theory had been transformed into a logical system by the time Condillac published his two main works, *Essai sur l'origine des connaissances humaines* (1746) and *Traité des sensations* (1754). For Locke, understanding was an "autonomous principle endowed with activity of its own"; for Condillac, in contrast, it was no more than "the sum total of a combination of psychic activities."[8] Judgment, desire, lust, and craving are nothing but modified expressions of sensation. And everyone remembers Condillac's proof of this: the fictitious statue that, in coming alive, confuses the sense of its own existence with the fragrance of a rose which it first smells.

Thenceforth all scientists, all philosophers have found themselves obliged to engage with sensualism. However great their resistance, they have been unable to escape its influence.[9] These, however, are mere episodes in the history of philosophy during the Enlightenment. The important point here concerns the developing alertness to the sensory environment. The senses "increasingly [became] analytical tools, sensitive gauges for the degree of pleasantness or unpleasantness of the physical environment."[10] While Hallé with his sensitive nose pursued the menace of germs, an optimist like Abbé Pluche invited his readers to enjoy the spectacle of Nature.[11]

The philosophers, however, paid little attention to the sense of smell. This neglect reinforces Lucien Febvre's argument that the sense of smell has declined in importance since the beginning of the modern period.[12] Moreover, scientific discourse has been reluctant to address this issue, given the extent to which it is riddled with contradictions; science has oscillated between appreciating and depreciating olfactory phenomena. The baffling poverty of the language,[13] lack of understanding of the nature of odors, and the refusal of some scientists to abandon the *spiritus rector* ("guiding spirit") theory all help to explain the abundance of muddled thinking and tortuous writing on the subject.[14] A few fairly simple stereotypes demonstrate the paradoxical nature of the sense of smell. Olfaction as the sense of lust, desire, and impulsiveness is associated with sensuality. Smelling and sniffing are associated with animal behavior.[15] If olfaction were his most important sense, man's linguistic incapacity to describe olfactory sensations would turn him into a creature tied to his environment.[16] Because they are ephemeral, olfactory sensations can never provide a persistent stimulus of thought. Thus the development of the sense of smell seems to be inversely related to the development of intelligence.

Unlike the senses of hearing and sight, valued on the basis of a perpetually repeated Platonic prejudice, olfaction is also relatively useless in civilized society. According to Count Albrecht von Haller, "The sense of smell was less important to [man], for he was destined to walk upright; he was to discover from a distance what might be his food; social life and language were designed to enlighten him about the properties of the things that appeared to him to be edible."[17] The best proof of this claim is that the sense of smell is more highly developed among savages than among civilized men; Père du Tertre, Père Lafitau, Humboldt,[18] Cook, and the early anthropologists[19] were agreed on this point. Even if some anecdotes on this subject seem exaggerated, observations of adolescent savages have confirmed that people who have been raised outside civilized society have a superior sense of smell.[20]

All these scientific convictions produced a whole array of taboos on the use of the sense of smell. Sniffing and smelling, a predilection for powerful animal odors, the erotic effect of sexual odors—all become objects of suspicion. Such interests, thought to be essentially savage, attest to a proximity to animals, a lack of refinement, and ignorance of good manners. In short, they reveal a basic failure at the level of social education. The sense of smell is at the bottom of the hierarchy of senses, along with the sense of touch. Furthermore, Kant disqualified it aesthetically.

Jean-Noël Hallé's behavior contradicts all these assertions and demonstrates the first paradox: the sense of smell is an animal sense— and at the same time, and precisely because of this, the sense of self-preservation. The nose, as the vanguard of the sense of taste, warns us against poisonous substances.[21] Even more important, the sense of smell locates hidden dangers in the atmosphere. Its capacity to test the properties of air is unmatched. The increased importance attributed to the phenomenon of air by chemistry and medical theories of infection put a brake on the declining attention to the sense of smell. The nose anticipates dangers; it recognizes from a distance both harmful mold and the presence of miasmas. It is repelled by what is in a state of decomposition. Increased recognition of the importance of the air led to increased acknowledgment of the importance of the sense of smell as an instrument of vigilance. That vigilance produced the guidelines for the reordering of space when the rise of modern chemistry made that reordering unavoidable.

A second paradox is that olfactory sensations are ephemeral, and thus defy comparisons through memory; any attempt to train the sense of smell always results in disappointment. This is why olfaction would not be taken into account in designing the English garden as the privileged place of sensory education and fulfillment.

On the other hand, doctors since ancient times have untiringly stressed the importance of the nose as the sensory organ closest to the brain, the "origin of sensation."[22] Moreover, "all the fine threads of the olfactory nerves and plates are extremely loose and full of life. Those that are more removed from this source are firmer and less penetrable, in line with the general laws on nerves."[23] Thus, in contrast to the claims made in association with the first paradox, the extraordinary subtlety of the sense of smell appears to *grow* with the development of intelligence. The culminating aesthetic proof is that the exquisite fragrance of flowers "appears to be made for man alone."[24]

As the sense of affective behavior and its secrets (in Rousseau's frame of reference, the sense of imagination and of desire),[25] the sense of smell was viewed as capable of shaking man's inner life more profoundly than were the senses of hearing or of sight. It seemed to reach to the roots of life.[26] In the nineteenth century it was elevated to being the privileged instrument of recollection, that which reveals the coexistence of the self and the universe, and, finally, the precondition of intimacy. Furthermore, the rise of narcissism was also to favor this hitherto discredited form of sensory life, in the same way that the obsessive fear of polluted air and the battle against infectious disease emphasized its importance.[27]

Clearly, the theoretical discourse devoted to olfaction reflects a maze of fascinating taboos and mysterious attractions. The required vigilance toward the threat of putrid miasmas, the exquisite enjoyment of fragrant flowers and the perfume of Narcissus counterbalance the proscription of sensuous animal instinct. It would be therefore an overhasty move to exclude the sense of smell from the history of sensory perceptions simply because of the infatuation with the prestige of sight and hearing. The following chapters investigate the modes of behavior that were caught up in the vague and contradictory theories formulated about olfaction. Let us therefore begin by returning to the trail opened up by Jean-Noël Hallé.

PART ONE

The Perceptual Revolution, or the Sense of Smell on Trial

I *Air and the Threat of the Putrid*

UNTIL ABOUT 1750, before the crucial advance in what was called pneumatic chemistry, air was regarded as an elementary fluid and not as the product of a chemical combination.[1]

After the publication of Stephen Hales's books, however, scientists had become convinced that it entered into the very texture of living organisms. All the mixtures that composed the body, fluids as well as solids, gave off air when their cohesion broke down. This discovery extended the presumed field of action of air. Henceforth it was thought to act on the living body in multiple ways: by simple contact with the skin or pulmonary membranes, by exchanges through the pores, and by direct or indirect ingestion (since foodstuffs also contained a proportion of air, which could be absorbed into the chyle and hence into the blood).

By its physical qualities, which varied according to region and season, air was thought to regulate the expansion of the fluids and the tension of the fibers. Once its weight was accepted as a scientific truth, scientists acknowledged that it exercised a certain pressure on organisms. Thus life would be impossible unless an equilibrium were established between the external and the internal air; this equilibrium was precarious, continually reestablished by belching, breaking wind, and ingestion and inhalation.[2]

Air was easily compressible and was activated by a mechanism equal in force to gravity. The smallest air bubble held in balance the weight of the atmosphere. This force made respiration possible, maintained internal movements, ensured expansion to compensate for the constriction caused by the weight of the fluid. Air never lost its elasticity of its own accord; but once lost it was not automatically regained. Only movement, agitation, permitted the elasticity of the atmosphere to be restored and organisms to survive. Death occurred when the fluid no longer had the force to enter the lung.

The temperature and humidity of air exerted an influence on bodies. By a subtle interaction of contraction and expansion, they helped to break down or restore the tenuous equilibrium between the internal environment and the external atmosphere. Heat tended to rarefy the air, and thus caused a slackening and an elongation of the fibers. The external parts of the body, especially the extremities, swelled. The whole organism experienced weakness, even prostration. Cold air, on the other hand, contracted the solids, compressed the fibers, condensed the fluids, and increased the individual's vigor and activity.[3] Paradoxically, people continued to believe that it was air which cooled the blood and thereby regulated the perceptible as well as the insensible perspiration shown by Sanctorius in the seventeenth century.[4] On the one hand, fresh air was therefore particularly beneficial.[5] On the other hand, there was a risk that overcold air would hamper evaporation of the excreta and cause scurvy.

Heavy humidity, morning or evening dew, and persistent rain relaxed the solids and stretched the fibers, because they helped the fluid to enter through the pores and simultaneously weakened the elasticity of the internal air. By a combination of these noxious effects, hot and humid air might jeopardize the precarious equilibrium necessary for survival.

Air also acted as a passive carrier, transporting an accumulation of foreign particles.[6] Like its physical qualities, the load carried by this heterogeneous fluid varied with time and place and were correspondingly described by individual theories far too numerous to recount here. But most experts were agreed in regarding air as the substance in which Stahl's phlogiston expanded and, for that reason alone, as indispensable to life. It was also seen as the vehicle of heat. According to Boissier de Sauvages, air ensured the transmission of electric fluid that maintained its elasticity.[7] Many thinkers attributed to air the transmission of magnetic particles, even of vague astral influences.[8]

On the other hand, no one at that time doubted that air also held in a state of suspension the substances given off by bodies. The atmosphere was loaded with emanations from the earth and with the perspirations of vegetable and animal substances. The air of a place was a frightening mixture of the smoke, sulfurs, and aqueous, volatile, oily, and saline vapors that the earth gave off and, occasionally, the explosive material that it emitted, the stinking exhalations that emerged from the swamps, minute insects and their eggs, spermatic animalcules, and, far worse, the contagious miasmas that rose from decomposing bodies.

It was this unfathomable mixture whose components Boyle strove—with no great success—to analyze by means of improvised methods.[9] This seething mixture, ceaselessly altered by various kinds of disturbances, was a theater of strange fermentations and transmutations in thunder and lightning, reconstituted by storms when the superabundant sulfuric particles were removed. It was also a menace in very calm weather, when deadly stagnation transformed sheltered ports and deep bays into sailors' graveyards.

As the physical properties of air acted collectively and individually, so the composition of its contents governed the health of organisms. Sulfur, stinking emanations, and noxious vapors threatened its elasticity and posed threats of asphyxia; metallic acid salts coagulated the blood of the capillary vessels; emanations and miasmas infected the air, incubated epidemics. Rooted in this body of beliefs was the atmospheric vigilance that underlay neo-Hippocratic medicine; in the fifth and fourth centuries B.C. Hippocrates and his disciples at Kos had already emphasized the influence of air and place on fetal development, the formation of temperaments, the birth of passions, the forms of language, and the spirit of nations[10]—ideas which were to give rise to epidemiology in the latter years of the ancien régime, and which inspired the "pneumatopathological" table formulated by the Société Royale de Médecine.[11]

"Every animal is adapted to the use of fresh, natural and free air," wrote Arbuthnot; young animals lacked that tolerance, born of habit, which allowed the city dweller to withstand "artificial air."[12] Therefore, even before researchers such as Priestley or Lavoisier set out to analyze "common air," breathing air that was not loaded with a noxious burden was being claimed as a natural right; only later did the idea of purity become linked to that of decomposition. For the moment, the focus was on the correct equilibrium between the "tainted" and the "purified,"[13] an impossible quest that ordained a private hygiene based entirely on distrust of changes in the climate, sudden thaws, bursts of mild wet weather, or downpours after extensive drought. This discourse on hygiene placed a high symbolic value on whiteness of complexion and transparency of skin, as obvious signs of the quality of the aeriform exchanges on which the living being thrived.[14]

Definitions of the healthy and unhealthy were sketched out and norms established in terms of aerist thought. The need for ventilation was already being formulated, and the poetic hymn to the purging power of the storm was taking shape.

The tentative discoveries made from 1760 until Lavoisier formally and publicly identified breathing with combustion in 1783 profoundly changed pneumatic chemistry. Before 1760, the sense of smell was not closely involved in assessments of the air; it was a long way from generating the anxiety linked to the advance of aerism. The physical properties of the atmosphere were measured by touch or by scientific instruments, and various factors—such as the theoretical nature of the discourse on miasmas and viruses, the vagueness of emanations, the absence of accurate analyses, and the imprecision of a vocabulary in its formative stages—contributed to downgrading the sense of smell. It is significant that references to smells in the contemporary debate between supporters and opponents of contagion were rare.[15]

Between 1760 and 1780 chemists did their utmost to achieve some precision in the analysis of the threat.[16] They formulated a twofold project: first, to establish a list and give names to the mixtures, while striving to create a language that would make it possible to define them by smell; second, to identify the stages and rhythms of corruption and to situate them on a scale based essentially on smell, since olfaction was confirmed as the sense most appropriate for observing the phenomena of fermentation and putrefaction. The stumbling beginnings of eudiometry did little to check the increased scientific role of the nose, which, though not a precise tool of analysis, was infinitely more sensitive than the apparatus perfected by Volta or Abbé Fontana.

From the 1770s on, chemists and doctors refined the vocabulary that would permit them to transcribe observations of the sense of smell. The translation of olfactory vigilance into scientific language led to a striking increase in references to smells by all the experts in the late eighteenth century. Henceforth this vigilance had manifold

aims: to detect irrespirable gases and particularly "airs," and to discern and describe hitherto imperceptible viruses, miasmas, and poisons; an impossible project based on error but one that doctors
pursued until Pasteur's theories triumphed. While pioneers in clinical
medicine were establishing a framework for understanding disease
by studying the lesions of corpses, most medical researchers, combining neo-Hippocratism and the mechanistic tradition, accounted
for smells arising from pathological cases on the scale determined
by observation of putrid decomposition.

Between 1770 and 1780 scientists passionately collected, de-
canted, confined, and preserved "airs"—also called gases—and located the effects of each on the animal organism.[17] In the space of
a few years they had drawn up a table of respirable "airs" and stinking
emanations. It was a confused, tangled classification, and the ter-
minology was still ambiguous. A few leading elements were singled
out: fixed air, sulfuric acid, inflammable air, volatile alkali, and liver
of sulfur. During these multifarious experiments, researchers learned
to track and recognize the various members of this prolific family by
means of the sense of smell.

While rats, dogs, and rabbits writhed and died in bell jars, the
interactions and transmutations linked to the vital mechanism of life
were gradually revealed. Joseph Priestley measured the deterioration
of "common air" used in respiration and the production of "phlogisticated air" (azote) and "fixed air" (carbonic acid) at the expense
of dephlogisticated "vital air" (what we now call oxygen).[18] The last
was henceforth recognized as respirable air par excellence. However,
the British scientist's loyalty to phlogiston prevented him from pur-
suing his analysis to its correct conclusion. Priestley also outlined
the theory of the gaseous exchanges of the vegetable kingdom but
left Ingenhousz the credit of giving a precise description of photosynthesis. The discovery that plants generated oxygen when exposed
to light gave these two researchers an optimistic vision of a providential design that caused vegetation to correct the air that animals
corrupted.[19] As a result of these discoveries air was no longer regarded as an element but as a mixture of gases; the respective proportions of these gases determined its properties.

Priestley also showed that it was possible to calculate the rate of
the "respirability" of air. After that Abbé Fontana traveled through
Europe equipped with his eudiometer, claiming to be able to pro-
nounce on the vitality of the atmosphere. Disappointingly, it turned

out that there was not the slightest difference between the properties of air in the Les Halles district of Paris and those of air in the mountains; the hopes placed in the apparatus had very quickly to be abandoned. In the last resort only the sense of smell could divine the true meaning of these ominous emanations.

The Odors of Corruption

It was nevertheless the basic objective of pneumatic chemistry, which produced so many observations of organic phenomena, to eliminate "the vagueness of the putrid," and so produce a full understanding of the mechanisms of infection.[20] To study "airs" was to study the mechanisms of life; the fashion for the pneumatic experiment spread rapidly through enlightened circles. Air was no longer studied as the area of generation or of the burgeoning of vitality, but as the laboratory of decomposition.[21] This fascination was prompted by observations of the death throes, the breakdown of the parts of the living body. Researchers literally smelled out the progress of organic dissolution in order to map the escape routes of the body's "cement," fixed air, and thus to trace the breakdown of the cohesion of the body's essential mixtures.[22] It was a matter of following death's advance into living matter in order to detect how the mysterious vital equilibrium was established.

Francis Bacon might seem to be the originator of this research into decomposition.[23] At the beginning of the seventeenth century he was already proclaiming that all malfunctions of the human organism led to a greater or lesser degree toward putrid dissolution, which entirely destroyed the arrangement of the parts in order to permit a new combination. This phenomenon, he said, was proved by changes in bodily smells; for example, he had already observed, in the wake of putrefaction, the smells of ambergris, musk, and civet, which were often its products.[24]

But the real founder in modern times of the putrefaction theory was Johann Becher (1635–1682). He viewed decomposition as a continual internal movement perpetually in conflict with the principle of the natural and fiery cohesion of the parts, an elemental fire that was perpetuated by the balsamic (in this context, oily) spirit of the blood. In a mechanistic perspective, this internal movement resulted from the mobility of the molecules liberated from the shackles that held them in place. The fetid, penetrating odor of the rotting body arose from this mobility and therefore could not be regarded as a simple symptom of decomposition; it was an integral part of the process. Fetidity and humidity revealed and defined corruption. The aqueous parts of organic matter were liberated in the form of sanies and pus; the putrid parts became volatile and escaped in the form of foul-smelling molecules. The earthy parts remained. Gangrene, syphilis, scurvy, and pestilential or putrid fevers could triumph in the unceasing conflict enacted in the living being if putrid miasmas emanating from diseased or decomposing bodies were accidentally inhaled by the organism and broke the equilibrium of its internal forces, or if obstruction of the vessels, viscosity of the humors, or injury interrupted the circulation of the balsamic spirit of the blood.

An antiseptic—that is, a substance capable of controlling the excesses of putrefaction—was to be sought among volatile, hot, oily, aromatic bodies, which could prepare the paths through which the balsamic vital spirit must circulate. The therapeutic use of aromatics, based on their volatility and powers of penetration, confirmed that ancient tradition which had led Hippocrates to control plague by odors.[25]

Becher's natural philosophy led, as we shall see, to a twofold increase in the value placed on odors: on the one hand, the idea that fetidity reflected disorganization; on the other, the recognition that aromatics opened the way to the vital principle. Both symptoms and remedy derived from olfaction.

Taking up Boyle's statement that the corruption of organic matter produced air, Hales set out to study and measure that emanation. A cubic inch of pig's blood produced thirty-three cubic inches of an "air" that Joseph Black later christened "fixed air." After that, studies of putrid dissolution changed direction. Decomposition was considered a lysis, resulting from an intestinal movement. Thus scurvy (henceforth regarded as the paradigmatic putrid illness) was in fact a form of decomposition with its origins in the living body. But what ensured cohesion, what remained after the parts had broken up, was not terrestrial; it was air. The body's cement was volatile in nature; after it had left the body, its components—saline, oily, aqueous— entered into other combinations.

These major intuitions inspired the experiments by John Pringle in 1750 and, a few years later, by the Dublin chemist David MacBride.[26] Antiseptic, according to the latter, had to serve a quadruple function:

to stem the liberation of fixed air, which could result from the dissolution of the blood or from too great a relaxation of the fibers; to ensure the fluidity necessary for all internal movement; to facilitate expulsion of putrid matter in transit in the body; and, when necessary, to restore the decayed substances to their natural state. Pringle and MacBride therefore included in their list of antiseptics astringents that contracted the fibers as well as aromatics, salts, quinine, and, in the last resort, air itself.

These British discoveries spread rapidly in France. In 1763 the Dijon Academy set a competition on the study of antiseptics, which Boissieu won. He submitted a brilliant synthesis that stressed both the forces of putrefaction indispensable to all living organisms, and the precariousness of the vital equilibrium, which needed to be kept under continual supervision. He pinpointed the dangers and listed the principles and tactics that would guide later action by sanitary reformers. It was most important to contain the leakage of fixed air, since it tended to escape when there was no opposing force and to reenter the cycle of aeriform exchanges that governed life and death. To this end, certain hazards had to be avoided: first, heat, which tended to deplete the particles that composed the body and thereby weaken the protective system; second, humidity, which loosened the cohesion of the parts; third, immersion in air that had lost its elasticity and therefore offered less effective resistance to the escape of "fixed air." But what had to be avoided above all else was air infected by putrid exhalations, which transmitted to the liquids "the internal motion that stirred them up" and thereby hastened the advance of putrefaction.[27]

By contrast, the doctor should encourage everything that hampered the escape of the gas. He should ensure the movement of the fluids, because this kept the air in a state of fixity; see that excretion, which expelled the putrescent humors, functioned smoothly; facilitate the absorption of air by lungs, pores, and the inhalant vessels of stomach and intestines; improve gaseous exchanges through the intermediary of the chyle. He should do this by choice of foodstuffs, use of balsamic antiseptics, or exposure to the vapors escaping from heated aromatics or certain substances in a state of fermentation. In this way Boissieu defined a public health policy that went far beyond neo-Hippocratic theory. It was based on analysis of the air, the struggle against putrid miasmas, and the increased value placed on aromatics.

With the spread of such theories, experiments with smells and analyses of the odors of putrefaction proliferated. Becher himself had attempted to describe the odors of the different stages of putrid decomposition. In a thesis defended at Montpellier in 1760 Féou refined Becher's analysis. In the first moments of death a "sweetish odor" appeared, which some people regarded as "vinous fermentation." Then a stronger acidic odor developed, "quite often similar to that of decaying cheese"; Gardane described it as "acidocaseous." "Finally the odor of decay appears; at first it is stale, without pungency, but this staleness is nauseating . . . imperceptibly the odor becomes penetrating, then it is bitter and abominable. The putrid taste is followed by a herbaceous taste and the odour of amber." Féou concluded: "This must lead doctors to determine with greater precision the role of odors in diseases."[28]

The case of Madame Thiroux d'Arconville provides a fine example of the scientist ceaselessly alert for the smells of decomposition. Robert Mauzi has stressed the importance of this aristocrat who was passionately interested in natural philosophy.[29] Confined to the virtuous life by the stigmata of smallpox, she seems to have found compensation in the pleasure she derived from science.

She claimed to have carried out experiments on more than three hundred substances in order to study how decomposition could be controlled in each. The result was a six-hundred-page book, not counting illustrations.[30] The young woman took care to repeat experiments in relation to season, temperature, degree of humidity, winds, and exposure. She worked in both town and country and kept a record of all her scientific activity. Madame Thiroux d'Arconville was an incomparable observer of odors. Watching the incessant rhythm of putrefaction for months on end, she was overwhelmed by the variations in smell, bearing witness to immense mysteries. She found this variety more stimulating to the imagination than the change of color in decomposing substances or the hissings and bubblings of fermentation.

She was not alone in this fascination. Godart, who like Boissieu submitted a dissertation to the Dijon competition, confessed himself hypnotized by the discontinuous rhythms of the smells of decomposition and by what he christened the "deflagrations" of odors inside his jars.[31] Another example was Dr. Raymond, whose book on elephantiasis described how he set out to follow the progress of decomposition in the living body by means of his nose.[32]

The sense of smell benefited from the movement that, under the influence of the disciples of Locke and then of Condillac, progressively refined research into sensory phenomena and the analytic capacity of each of the senses. Contrary to a generally accepted idea, olfaction probably gained more from the movement than did sight, hearing, or touch. It was in fact more closely involved in the contemporary definition of what was healthy and what unhealthy, which helped govern public health behavior until Pasteur's discoveries. Whereas nascent clinical studies favored sight, hearing, and touch, the sense of smell could discern a subterranean physiology, changes in the humors, and "the order of putrefaction."[33]

To translate the new requirements, the vocabulary became more complex; using a whole system of indications conveyed by smells, the doctor had to know how to untangle a multitude of symptoms. Thanks to this apprenticeship he then knew how to handle a dual series of olfactory data: those that enabled him to recognize gases and thus detect the threat of mephitism; and those that related to analyses of fermentation and decomposition, and consequently made it possible to predict miasmas and locate their effects on the organism. The innumerable references to odor in subsequent medical writing and in Hallé's everyday experience are hardly cause for surprise.

But even though doctors and public health experts played a considerable role in transmitting this revolution in perception, everything suggests that they merely expressed their contemporaries' particularly acute sensitivity to smell. The sense of smell temporarily enjoyed an obvious increase in importance because, more than the other senses, it created those "new anxieties" that imbued the pre-Pasteurian mythologies.[34] Crucially, it was able to reveal the precarious nature of organic life. This concern with the smells of putrefaction opens up unfathomable perspectives on the psychology of the elite in the last days of the ancien régime. That constant monitoring of the advance of death in the living being, which spawned a careful analysis of belches, rumblings, wind, colics, and fetid diarrhea, introduced new anxieties. The calculation of degrees of internal decay based on the odor of bodily waste led to the astonishing excremental vigilance discussed in the next chapter.

Man's relationship to his environment was also revolutionized. The main considerations were no longer altitude, exposure, the quality of space, and the nature of the winds, but the qualities of the confined, enclosed area of everyday life, the aerial envelope and the atmosphere of bodies. The dangers henceforth were those deriving from "degenerated air," mephitism, proximity to foul-smelling substances, putrid molecules emanating from corruption, and the "aerial miasma," with its considerable power to threaten living matter; miasma's corruptive powers extended to plant life, butcher's meat on the slab, and cutlery.[35]

Concern with the smell of the putrid expressed the dilemma of the individual who was unable to *fix*—that is the crucial word—or to control the elements which composed him, which he had inherited from preceding individuals, and which would permit the combination of new individuals. Putrefaction was a time machine. Hence, from then on olfactory vigilance not only aimed to detect the threat, the risk of infection, but also entailed a permanent monitoring of the dissolution of individuals and the self. For us, as for Oscar Wilde's Dorian Gray, the marks of disintegration are visual; for Hallé's contemporaries, they were also odoriferous. Quite simply, foul-smelling miasma provoked panic.

Jacques Guillerme notes that Schlegel, among others, often put the putrid in the same category as the demoniac;[36] the effect was to strengthen the obsessive correlation between stench and the depths of hell, most apparent in authors who have tried to describe Gehenna, from Milton to Cowper Powys.[37] From a more limited historical viewpoint, anyone striving to understand the French Revolution would probably find it useful to put this fascination with putrefaction in line with contemporary fascination with corpses.[38] In any case, one historic fact of prime importance remains: the putrid was about "to become an image reflecting a social form of life as a typical form of nature."[39]

2 *The Extremes of Olfactory Vigilance*

IN STRESSING how much the old belief in the dangers of emanations from the earth still haunted scientific discourse during the first part of the eighteenth century, Jean Ehrard quoted Abbé Dubos on the subject: "By the action of the central fire, it [the earth] is subject to continual fermentations: thence the effluvia, which vary in nature with the nature of the subsoil; but as nothing is more unstable than a fermentation, they are scarcely less varied in time than in space."[1] In 1754 Boissier de Sauvages added more precise information: "A vapor rises from the whole surface of the earth as a result of subterranean heat—10° Réaumur. It is to a greater or lesser degree abundant, denser than air, and it spreads out when nothing prevents it, and falls again in the evening."[2] According to Pieter van Musschenbroek, four liters six ounces of this "transpiration of the earth" were deposited every year on every square foot of soil.[3]

The bowels of the earth were also the laboratory of a "physica subterranea"[4] that through mysterious mixtures was continually trying to counterbalance the deadly blasts by virtue of balsamic emanations.[5] The experience of miners was sufficient proof of the overwhelming noxiousness of some of these terrestrial vapors.[6] Ramazzini denounced the harmful effects of the distinctive odor of the mines.

Quarries also exuded terrible threats, notably the "metallic vapor which is exhaled by marble, tufa, and certain stones, and which obviously attacks the nostrils and brain"; even more dangerous was "the disagreeable odor" that rose from touchstone. Near Mount Zibinius, Ramazzini could discern from over a mile away the fetid emanations from the petroleum (rock oil) that infected the workers.[7]

Such observations revealed the dangers of agriculture and formed the basis of future discussions of the unhealthiness of the countryside.[8] In a report presented to the Société Royale de Médecine in 1786 Chamseru exposed the dangers incurred by the peasant who bent down and brought his face too close to the soil he was turning over.[9] Baumes demanded that agricultural workers be prevented from sleeping with their noses to the soil.[10] He deplored the fact that villages were constantly subject to "morbific vapors" released by cultivation. The perils were increased when an untimely land-clearing operation uncovered virgin soil; "how many colonies in the new world have been the unfortunate victims of the terrible fevers produced by deadly vapors from virgin and alluvial soil!"[11]

Even worse were the ravages caused by emanations from water-logged soil. In fact in a whole range of dangerous plots of land fermentation was incessantly at work. One example was the Maremma, near Volterra; the soil there was constantly threatened by "eruptive rocks," by "subterranean emanations," and by an "oleo-bituminous" substance; the saline lands exuded a gas that was unfit for respiration and poisonous miasma to which Savi a century later (1841) attributed intermittent fevers.[12]

All these beliefs engendered an obsession with fissures, faults, and imperfect joinings. Above all, it was important to keep watch on the boundaries of any dangerous terrain. These were the points of contact that allowed mephitic blasts of air to percolate.

The most terrible cracks were clearly those created by earthquakes. According to Tourtelle, the epidemics that ravaged Lisbon and Messina after these disasters had no other origin.[13] Reference to the danger of open fissures in the putrefied ooze of swamps became a commonplace. The subsoil of ponds exuded the worst stench. Fear of escaping emanations caused a dread of all faulty construction: cracked cesspools, badly joined floors, ill-secured paving, unsealed vaults and cisterns.

The earth not only spewed forth blasts of air; it also absorbed and stored the products of fermentation and putrefaction. It became a repository for foul substances. Someday it would send the morbific vapors back. There was an abiding obsession with subsoil impregnated, broken up, even liquefied by the accumulation of feces, the putrefaction of corpses, and the proliferation of fissures. Infected soil that had become foul-smelling was regarded as lost soil: future generations would not want to settle on it. Their excremental past weighed like a nightmare on the future fate of these places. The remains, waste, and excavations of past generations were seen as producing undeniable stenches believed to corrupt or disequilibrate the living organism. All these obsessions were strengthened by the survival, conscious or unconscious, of collective beliefs about the living nature of the earth.[14]

There were places where impregnation was extreme, the stench intolerable, the threat imminent. Near the refuse dumps of Montfaucon there was already a danger of "streams forming under the earth that were large enough and continuous enough to infect the wells in the neighborhood and suburbs, and to damage the strata of the earth or the foundations of dwellings"; there was a risk that "fetid material from the sluices" would penetrate the soil and infect the sites of future buildings.[15] In 1780 Lavoisier reported on the work of a commission of the Académie Royale des Sciences, which had been instructed to inspect the Saint-Martin and For l'Evêque prisons. The whole "terrain that formed the base" of those establishments was "completely penetrated with stinking and putrid material . . . Such a mass of corruption necessarily emits a *continual stinking exhalation.*"[16] The foul-smelling prison had to be abandoned solely because of its past unhealthiness. The underground jail was a privileged locus for remembrance of the past; its graffiti served as a reminder of the passage of time and the succession of inmates. It was natural for dungeons and oubliettes to crystallize anxiety about the putridity accumulated in the memory of the soil.[17]

The campaign against emanations from corpses was directed not only against organic debris; it persistently attacked the impregnation of the earth by evil-smelling liquids.[18] On the eve of the Revolution, the whole capital seemed to be undermined, wide open to morbific influences from an undermined and unreliable subsoil. According to Bruno Fortier, in about 1740 this belief initiated the chain of anxieties that guided sanitary reformers.[19] The fear was even greater by the end of the ancien régime. A silent fermentation threatened to engulf the city. The houses "are leaning over abysses," protested

Louis-Sébastien Mercier.[20] There was a risk that fissured cesspools would cause fatal landslides; accidents were multiplying. At the time of the flood in 1740 a movement to analyze the Paris soil was already beginning; later, the St.-Jacques district and part of the faubourg St.-Germain were shored up.[21] Many private individuals condemned the presence of graves and dungeons. Many contemporaries imagined their relation to waste matter and putrefaction as that of hapless victims before a providential or historical ineluctability.

Thus it is easy to understand the special alertness to miasmas. Mud, or rather the vapors that rose from it, became the subject of an anxiety-laden discourse. The multiplicity of descriptions, the meticulousness of the analyses seem astonishing: material to delight Gaston Bachelard.[22] The mud of Paris formed a complex mixture of sand that had infiltrated between the paving stones, vile-smelling rubbish, stagnant water, and dung; carriage wheels kneaded it, spread it, spattered walls and passersby.[23]

The concern with mud persisted to the mid-nineteenth century. Parent-Duchâtelet placed at the top of his scale of stenches the odor of dishwater that had dried when thrown on the paving stones.[24] More significant was the long archaeology of miasmas on which the great chemist Chevreul, tireless collector and analyst of Paris mud, embarked in the mid-nineteenth century. For him, the healthiness of towns was a function of past deposits; organic material "sooner or later resulted in various sorts of infection."[25] He also ventured to analyze with his nose "the black ferruginous material to be found under the paving stones of Paris." He kept numerous specimens in stoppered glass flasks; for example, on December 20, 1846, he collected mud "taken from between and under the paving in the rue Mouffetard, near the pont aux Tripes."[26] He left the products to steep and refrained from sniffing them until December 20, 1852.

Distantly echoing old fears, Chevreul also condemned "the capillarity of mortar."[27] Walls, intended to separate and support, also proved to be conduits, sites of complex upward movements, and, like the soil, repositories for ancient filth. They combined deposits with the wafts of mephitisms and hence concealed a multiplicity of threats. Also fatal were emanations from new walls, the odor of plaster and humidity, which Piorry thought specific, though reminiscent of the smell of sulfur.[28] In Paris new buildings were left to prostitutes, a practice known as "drying out the plaster."[29] The gases emanating from a new wall caused neuralgias, acute joint or muscular complaints. The problem became the subject of medical debate in the nineteenth century.

Walls also preserved odors. At the St. Petersburg naval hospital, John Howard noted with satisfaction that patients were moved to different rooms in summer as a means of checking miasmic impregnation of the walls.[30] According to Philippe Passot, former prisoners recognized the odors of incarceration that impregnated the dungeon at Vincennes years after it had ceased to be a state prison.[31] This strange preservative power was deadly: "A doctor attended a woman afflicted with a gangrenous disease from which she died. Two years later, returning to the same place to visit another patient, he found the same gangrenous odor, the odor *sui generis*."[32] The walls had transmitted the disorder of the bodily tissues. The mephitism of walls and ceilings was sometimes astonishingly intense. After an epidemic of puerperal fever that claimed eighteen victims at the Lyons hospital, workers disinfected the murderous room; this task involved removing the layers of old mortar. "As they detached the rubble from the walls and ceilings, a most fetid odor was given off." Polinière, an expert, declared "that the stench was so great that it even surpassed the odor of a dissection room."[33]

Saltpeter caused a porous, greasy, humid coating to form on the walls between rooms, and this turned into a crust. Thenceforth emanations oozed continually from the walls. Our ancestors, wrote Géraud, used thick woolen tapestries as protection against these; to replace them with single layers of paper or cloth was a mistake.[34] The terror inspired by the incrustations and even by the moldy skins—seen as forming protective screens beneath which leavens proliferated and viruses were bred and swarmed—could form a serious topic of research on its own. The terror is discernible in numerous writings devoted to swamps, sewage, and building construction. Pouchet's proligerous skin, of symbolic significance, reflected this fascination.

Wood aroused the same type of anxiety. Lind, like Duhamel-Dumonceau, condemned the ravages caused by the smell of fresh woodwork that pervaded new ships.[35] John Howard was amazed at wood's capacity for absorbing smells: putrid emanations could reach to the heart of an oak trunk.[36] The floors of Worcester jail had been "rotted by . . . the prisoners' breath."[37] It was a commonplace that butchers' and fishmongers' stalls remained impregnated with the fetid odors of their merchandise. The complaints occur in all descriptions of markets.[38]

The Swamp of Excrement

There were less obscure odors than those formed to a greater or lesser degree by the complex fermentation of the earth, less ancient miasmas than those exhaled after slow impregnation. These more obvious threats, which aroused the vigilance of sanitary reformers, were the odors of excrement, corpses, and decaying carcasses. The strong smell of excrement pervaded the environment, and the stench of public places was both terrible and ceaselessly condemned. The vile-smelling effluvia of the faubourg St.-Marcel assailed the young Rousseau when he entered the capital. In the Palais de Justice, in the Louvre, in the Tuileries, at the Museum, even at the Opera, "one is pursued by the unpleasant odor and infection from the places of ease [latrines]."[39] In the gardens of the Palais Royal "one does not know where to sit in summer, without inhaling the odor of stagnant urine." The quays revolted the sense of smell; excrement was everywhere: in alleys, at the foot of milestones, in cabs.[40]

The cesspool clearers made the street stink;[41] to save themselves trips to the refuse dumps, they let the barrels empty into the gutter. The numerous police ordinances relating to this scourge were not enforced.[42] Fullers' and tanners' workshops helped spread excremental odors.[43] The walls of Paris houses were stained by urine. Louis-Sébastien Mercier waxed apocalyptic when he recalled the "amphitheater of latrines, perched on top of one another, adjoining stairways, next to doors, very near kitchens, and exuding the most fetid odor on all sides," or the frequency with which blocked pipes cracked, flooded the house, and blasted pestilence through stinking shafts that seemed like the mouths of hell to terrified children.[44] In short, Paris, "center of science, arts, fashion, and taste," stood out as "the center of stench."[45]

Versailles itself was no exception; the cesspool was next to the palace. "The unpleasant odors in the park, gardens, even the château, make one's gorge rise. The communicating passages, courtyards, buildings in the wings, corridors, are full of urine and feces; a pork butcher actually sticks and roasts his pigs at the bottom of the ministers' wing every morning; the avenue Saint-Cloud is covered with stagnant water and dead cats";[46] livestock defecated in the great gallery; the stench reached even the king's chamber. On the eve of the Revolution, Arthur Young drew a map of urban stenches; those at Rouen, Bordeaux, Pamiers, and, above all, Clermont choked him. In the capital of the Auvergne he found many streets "that can, for blackness, dirt, and ill scents, only be represented by narrow channels cut in a night dunghill."[47] In this and similar descriptions the salient feature was the emergence of a new sensitivity; we shall return to this phenomenon later.

Opinion on excrement was divided; its therapeutic value was contested in scientific circles. Although Pringle himself stressed the need to distinguish between the odor of feces and the threat of putrescence,[48] many writings stressed the danger of gases released from excrement.

On the eve of the Revolution, there were increased attempts to analyze the noxious gases that rose from cesspools, particularly when they were being cleared. The workers had to be protected against suffocation. This scientific work continued to be caught up with belief in the putrefying power of fecal odors, where the major danger was thought to lie. "The vapors from latrines," wrote Géraud, "corrupt all types of meat and its juices . . . this corruption takes place through the absorption of the 'air' of the meat by the putrid exhalations of latrines."[49] The clearing of cesspools also endangered the environment: "the air is corrupted by it, houses infected, the inhabitants inconvenienced, invalids endangered;"[50] flowers withered, young girls' complexions faded.[51]

There were various degrees of danger. The greatest came from excremental stagnation. It was of primary importance to avoid retention and thereby concentration of excremental matter. But this solution had been adopted in the capital as early as the edict of Villers-Cotterêts (1539). Thenceforth cesspools aroused keen anxiety. In the organicist perspective of the sanitary reformers, there was a risk that this social blockage would produce a chaos of filth in the city. Excrement was much more of a danger in town than in the country. Louis-Sébastien Mercier envied the peasants who relieved themselves in the fields while town dwellers risked putrid fever by sitting on deadly commodes.[52] Thouret noted that exposure to air and sunlight rendered the fecal matter spread out in the Montfaucon basins innocuous,[53] as was proved by the transmutations in the smells. If old excrement proved so dangerous, it was because it had become "alien to ourselves, our food, and our furnishings" by an interplay of "decompositions" and "recompositions."[54] It had lost the odor of the body. It had putrefied. Preserving the current system of retention would risk making "future races" pay dearly for this imprudence.[55]

So it is easy to see how excrement became a favorite topic of discussion, surprising though this may seem at first, given the fact that it broke the injunctions of the new Lasallean courtesy taught in the schools. But the need to impose silence on children on this subject clearly betrays their parents' anxieties. Excrement became a subject of conversation at the court of Louis XVI.[56] Voltaire commented that man had not been created in the image of God, since He did not have to satisfy such needs.[57] Mercier noted the spread of the habit of "looking into the bottom of the sluices."[58] Beaumarchais's *Parades* attested the extent of this fascination. Nougaret and Marchand even brought the cesspool clearer onto the stage.[59] Scientists carried out more and more analyses of smells. They tried to map the itinerary of excrement in the same way that Becher and his followers had set out to trace the stages of decomposition. One example will suffice. Hallé patiently listed the "exhalations and vapors . . . that emerge from cesspools," taking care to distinguish gases from "stinking effluvia" not cataloged by pneumatic chemistry.[60] His hierarchy of smells consisted of a spatial pigeonholing of odors—the odor of fresh excrement, the odor of the latrine, the odor expelled by pits, the odor of cesspool clearing—that corresponded to a progressive aging and corruption of fecal matter.

The scatological theme proved rich in implications. The phantasm of the excremental swamp, the horror aroused by the mishaps that befell cesspool clearers or individuals who drowned in their own cesspools, the terrible stories of lost travelers swallowed up at Montfaucon—all strengthened anxiety concerning the subsoil of Paris.[61] The stench and corruption from the accumulation of excrement challenged the city's very existence. From a quite different perspective, Louis-Sébastien Mercier set out to emphasize the egalitarianism of excrement, exposed to everyone's view, its stench submerging the entire capital. It was a constant reminder that everyone was reduced to the same condition in the act of defecation.[62]

The intensive treatment of social attitudes toward death by specialists in eighteenth-century history eliminates the need for lengthy explanations here.[63] However, death called for the highest degree of olfactory vigilance. Ever since the chemists had come to regard "fixed air" as the body's cement, death was seen as circulating in the atmosphere with the odor of corpses. Internal putrefaction cohabited with the vital principle inside organisms; the first was responsible for the abiding presence of death; and the gases and putrid emanations that rose out of corpses made it infiltrate the actual texture of the atmosphere. "Fixed air," greedy for new combinations, enveloped the living, threatened to upset the vital equilibrium, and made a mockery of the traditional barrier of the tomb.

After 1741, when Stephen Hales took up the subject, scientists encouraged a new alertness to the odor of the airborne remains of the dead. In 1745 Abbé Porée denounced the stench from tombs built in churches, though still only in terms of their unpleasantness to the senses.[64] A year earlier Haguenot attributed the accidents that occurred when sepulchers were opened both to the loss of the elasticity of the air and to the exhalation of putrid miasmas.[65] At the end of the century, Vicq d'Azyr abandoned study of the physical properties of the fluid released, and adopted a dual procedure that should by now be familiar to the reader. He outlined a chemical analysis of the gases that escaped from vaults, dismissing Priestley's "phlogistic air" and Volta's "inflammable air," arguing in favor of Black's "fixed air." He attributed the observed cases of suffocation to this unbreatheable gas but, like most of his contemporaries, continued to think that the principal danger lay in a "stinking vapor." Whereas gases "kill immediately, the second acts more slowly on the nervous system, as well as on the fluids of animals, which it manifestly impairs."[66] As de Horne stressed in 1788, the danger was all the greater in that these disturbances were often delayed and it therefore proved difficult to detect their cause.[67] Both vaults and natural cavities in the soil stored up these stinking vapors. This phenomenon explained the accidents that occurred in the cellars of shops bordering the cemetery of the Innocents.[68]

Although doctors were fully accustomed to handling and dissecting corpses, they were not exempt from anxiety. For example, when, in the course of a student examination in anatomy assigned to Nicolas Chambon de Montaux, by the dean of the Paris Faculty of Medicine, the liver from a putrefied corpse was used, the first of the four candidates,

> struck by the putrid emanations escaping from it immediately it was opened, fainted, was carried home, and died within seventy hours; another, the famous Fourcroy, was affected with burning eruptions on the skin; the other two, Laguerenne and Dufresnoy, languished for a long time; the latter never recovered.

As for Chambon, he was consumed with indignation at the dean's persistence and remained steadfastly in his place. He completed his lesson amid stewards drenching their handkerchiefs in aromatic water, and probably owed his health to that cerebral over-excitement which brought on a copious sweating during the night, after a few outbreaks of fever.[69]

Doctors since Becher had regarded the first emanations from the corpse as the most dangerous. For that reason alone, proximity to battlefields proved perilous in the extreme. Twenty years later, Fodéré set out to determine the thresholds of noxiousness.[70] Postulating that the range of action of putrid miasmas coincided with that of stinking emanations, he embarked on a system for measuring smells that enabled him to establish a spatial scale of threats of infection. All this scientific work (which we need not consider in detail here) underlay the many subsequent analyses of the smell of urban cemeteries and their charnel houses.[71]

The presence of slaughterhouses within towns promoted indignation and intensified vigilance concerning decaying carcasses.[72] The urban slaughterhouse was an amalgam of stenches. In butchers' narrow courtyards odors of dung, fresh refuse, and organic remains combined with foul-smelling gases escaping from intestines. Blood trickled out in the open air, ran down the streets, coated the paving stones with brownish glazes, and decomposed in the gaps. Because blood transmitted "fixed air," it was the most eminently putrescent of animal remains. The malodorous vapors that impregnated roadways and traders' stalls were some of the deadliest and the most revolting; they "make the whole body susceptible to putridity."[73] Often the stifling odors of melting tallow added to this foul-smelling potpourri.

Montfaucon remained the epicenter of stench in Paris. That malodorous complex, where sewage reservoirs and slaughterhouse stood side by side, took shape northeast of Paris during the second half of the century. It formed the first link in the (partly imaginary) girdle of putrid stenches that gradually encircled the city, cutting off all possibility of escape to less impregnated soils. The suspected underground waves of sewage and the scourge, blown by winds from the northwest, little by little produced the phantasm of a putrid tide beating at the gates of the capital. When he described this archetype of stench, Thouret waxed dithyrambic: "One has to traverse those places of infection to know what those residues or products that can be called the excrement of a great city really are and to understand what the immeasurable increase in uncleanliness, stench, and corruption that results from the proximity of men really looks like."[74]

Decaying carcasses and excrement initiated the cycle of impregnation and exudation of the soil, the dialogue of earth and air, that henceforth was thought to have the potential to transform the city into an infernal bog of waste matter.[75] And, however urgently the analysis of soil-smell galvanized the municipal authorities into action, we should not forget the extent to which fear of the soil continued to hold sway over the collective imagination.

Water inspired distrust even beyond any reference to smell. This fact helps explain the ways in which the strategy of deodorization was executed. Humidity in itself concealed a number of perils: it relaxed the fibers, thinned the humors, and therefore, according to Pringle, resulted in susceptibility to putrefaction.[76] In addition, water vapor was loaded with all manner of remains that were brought in with fog. Evening dew was noxious.[77] Washing with too much water involved dangers, notably to the vessels, which were areas of intense putridity. Saline vapors from the sea aroused particular distrust.

Any stagnant water was a threat. It was movement which purified. Currents expelled, crushed, dissolved the organic remains that lodged in the interstices of water particles; Hales carried out innumerable experiments on water from the Thames to demonstrate this phenomenon. Terrifying poisons gushed out of a barrel that had been left sealed for too long, or from a cistern that had been closed too tightly. "A sailor fell down dead unbunging a cask of seawater when the king's supply ship *Le Chameau* was laid off at the port of Rochefort; six of his mates who were some way away from him were struck down, racked by violent convulsions, and lost consciousness; the ship's doctor, who had hurried to their aid, experienced the same symptoms: the dead man bled from the mouth, nose, ears; his black and swollen corpse decomposed so rapidly that it was not possible to open it up."[78]

Even fresh water could exert terrible effects. The gardener at the hospital at Béziers was struck down by "mephitic gas . . . from water intended for watering the garden." This water was stagnant, black, thick and viscous, "still covered with a mossy and heterogeneous crust." The "murderous vapor was so effective that it struck this unfortunate gardener dead, although it was discharged in the open air, over a period of half an hour, and at a distance of a few *toises* from the reservoir . . . the following morning it was still strong enough to suffocate a young lay sister who courageously volunteered" to replace the "fatal" plug.[79] Abbé Bertholon, who reported the event, noted that the most fearful poison was involved, faster than a bullet, sharper than an arrow.

It is therefore understandable why these rivers of stench (of which the Bièvre in Paris, as the site of confluence of organic remains, was the symbol) remained for so long a focus of anxiety. Proximity to water increased the harmful effects of fermentation and decomposition; desiccation in the open air lessened the danger. The sun created the upward movement that brought deliverance; humidity, in contrast, forced the heavily laden miasmas to the ground. Thus the source of the greatest fear and danger was totally unrelated to what we, from a Pasteurian viewpoint, would regard as pollution.[80] Fourcroy and Hallé would have agreed that the excrement and refuse tipped into and dissolved in the Seine did not impair its purity. The real peril lay in the decay of carcasses downstream, in decomposition along the flat and muddy banks, in the exposure of remains that the current ceaselessly set down and picked up.

The area of stagnation and accumulation par excellence was the "marsh"—a diffuse concept that scientists had been striving to define since Lancisi. Even the smallest puddle was a threat; hence advice against inopportune washing. The depressions in the gaps between ill-joined paving stones were so many small bogs. The ravages caused by stagnant water in urban ditches or in the ponds that formed more or less spontaneously in the countryside, prompted incessant complaints. The foulest-smelling water was the most dangerous; the worst were those reservoirs, ponds, or "fisheries" in which hemp was steeped.

A whole cosmology was developed around marshes. Vegetable remains in a state of fermentation, decomposed organic waste, and the corpses of all the unspeakably foul creatures produced by organic decomposition mingled in the vile-smelling ooze. Vapors mingled incessantly in the movements between the subsoil, the fetid turf that covered it, and the aquatic mass. The hidden cycles of subterranean life unfolded under the crust or skin that formed a veil upon the surface of the liquid. Analysis revealed a life of which sight had no intimations but which stench betrayed. "When water from ponds or swamps was evaporated over a low flame, several worms, several insects, and other animals were deposited, along with a great deal of yellowish, earthy material." This water was "overloaded with substances that were foreign to it, emanations, vapors, exhalations from earth, mines, ooze, plants, fish, decomposed insects, and other substances with which the air is always to a greater or lesser degree infected."[81]

The most deadly swamps were those like the abandoned salt marsh of the Charente coast, where fresh and salt water mingled, "Either because the sea washes up and deposits a larger number of fish and insects, which die and decompose there, or because this mixture of fresh and salt water is well suited . . . to hasten the decay of the vegetable and animal organic molecules that water, even rain water, almost always contains."[82]

The danger became overwhelming when the veil was lifted from this swarming cloacal mass. It was always perilous to lay bare the soil; even mowing the vegetation, and hence uncovering the humid vapors locked away in vegetation, was to release effluvia.[83] But clearing an urban ditch or carrying out ill-considered drainage was courting epidemic. Plowing recently drained soil was equivalent to suicide. Particularly suspect were plots from which rivers had receded after flooding, especially when these plots were muddy and when the waters retreated in summer. The terrible dilemma was that, as soon as the odor drew attention to the dangers, the swamp had to be destroyed and the meanders of the rivers drained.

Chemists were happily engaged for nearly half a century in observing gases along marsh banks.[84] Once Louis XV's doctor, Chirac, had related the harmful effects of marshes to the smells they exuded, scientists were encouraged to use their sense of smell. Here the exhalations wove a "silken thread" over the water and stirred it up with sinister bubblings.[85] On the edges of swamps the effluvia could not be doubted. Sight and hearing confirmed the sense of smell. "A fetid and sometimes intolerable odor proclaimed the virulent nature of the exhalations that followed ceaselessly one upon another . . . Those who have learned to assess the nature of this odor relate it to that of tansy or gunpowder; some say it resembles that of cadavers."[86] Fodéré, who stayed at Valençay, was able to study "the evil stench" of the Brenne at his leisure.[87]

As on riverbanks, exhalations were held to be more dangerous in the evening, when they stagnated or returned from the atmosphere, than when the sun was drawing them up; however, it was in broad daylight that they appeared most intolerable. Observations of rhythms of smells during the daytime led Baumes to be cautious about equating the malodorous and the unwholesome.[88] This precaution was also valid for bodily effluvia. Everything in this context conspired to lead olfactory vigilance astray.

3 *Social Emanations*

The Odor of Bodies

EVERY ANIMAL SPECIES, every individual organism, Withof declared in 1756, had its distinctive odor. This oft-repeated statement was later the subject of copious comment by Théophile de Bordeu, an expert on glandular systems. Thus a belief inherited from the early days of science was assumed to be fact by medical scientists at the end of the eighteenth century.

Three major beliefs governed vitalist thought on the subject.[1] Bordeu defined them clearly. First, "each organic part of the living body has its fashion of being, acting, feeling, and moving: each has its taste, its structure, its internal and external form, its smell, its weight, and its fashion of growing." Second, each organ "does not fail to spread exhalations, an odor, emanations around itself, in its *atmosphere,* in its own province. These emanations have taken on its style and its demeanor; they are, in fact, genuine parts of itself . . . The liver colors everything surrounding it with its bile"; the flesh adjacent to the kidneys, for example, exhaled a vinous odor. Third, the humors were positive laboratories, continually transporting a strong-smelling "excremental vapor," which was evidence of the purging, the incessant repairing of the organism. This purging was achieved by the elimination of all the waste products: putrid effluvia, products of menstruation, sweat, urine, and fecal matter. The organism's "excretory ducts were always smoking."[2]

These and similar beliefs influenced medical science for over a century. Expanded at length by Brieude, Virey, and Landré-Beauvais, they were the basis of the works that marked the golden age of osphresiology (the science of smells), particularly Dr. Hippolyte Cloquet's *Traité des odeurs, du sens et des organes de l'olfaction,* published in 1821.[3] Twenty-four years later, Falize brought them up to date;[4] in 1885 Monin devoted a large, heavily documented work to the odors of the human body.[5]

The smells of the humors proved particularly amenable to interpretation. Barruel distinguished between male and female blood by odor.[6] According to Bordeu, menstrual products emitted a specific smell that enabled mothers to watch over their daughters' physiology because "there is a large quantity of invisible emanations in menstrual excretion."[7] This menstrual discharge should not be attributed simply to a plethora of blood, as hydraulicians claimed; it was part of the purging of the humors. This theory gave new impetus to the belief in the putrefying power of the menstrual flow, capable of spoiling sauces or the meat in the salting-tub. Yvonne Verdier has shown how it persisted in the village of Minot.[8]

Bile, too, engendered stenches by its putrefying effect. Milk impregnated the particular atmosphere of women. According to Bordeu, milk was subject to a process of ceaseless ebb and flow: "our women sweat milk, urinate milk, chew and sneeze milk, and they pass it in their stools."[9] Milk could even flood the womb.

Nevertheless, the decisive role was played by sperm, a "typical" liquid on which all the other humors were modeled.[10] Semen by definition formed the essence of life. It exercised its effect on the totality of the organism; its odor betokened the individual's animality. According to Withof, the seminal humor "nurtured" the male organs and stimulated all the fibers; it produced "that fetid odor which vigorous males exude" and which eunuchs lacked.[11] In man, this *aura seminalis* ensured the link, formed the connection between body and soul.[12] The "unsubtle" odor of the pubescent male, arising from the discharge or flowback of semen into the blood and organs, should not, it was maintained, cause disgust. Moreover, Brieude emphasized that, unlike the odor of the other humors, this odor never varied.[13] The theory that the tissues were impregnated with odors—a theory backed by the authority of Albrecht von Haller—was reexamined at length in the nineteenth century.[14] The seminal odor of the continent priest or of the assistant schoolmaster, the unsavory celibate, became a commonplace of Romantic literature; Jules Vallès denounced it as late as 1879.

The odors of the organs and of the humors, loaded to a greater or lesser degree with the products of purgation, were exhaled by the excretory ducts.[15] According to Bordeu, there were seven of these, all remarkable for their strong odor: "the hairy area of the head, the armpits, intestines, bladder, spermatic passages, groins, the gaps between the toes."[16] Strong-smelling effluvia were a sign of intense animalization and evidence of the vigor of the individual and the race.[17] Thus it was discovered that very ancient therapeutic practices had a scientific basis. The cure for any ailment arising from insufficient animalization was traditionally sought in stables containing young animals. When David was growing old, the presence of young naked girls in his bed restored his vigor. Capivaccio used the same cure for a young and languishing aristocrat, and Boerhaave for a German prince in a condition of decline. Some elderly schoolmasters were convinced that the air emanating from children's bodies had beneficial effects.[18]

Such beliefs led to considerable inhibition regarding personal hygiene. Both ethnologists and historians repeatedly stressed the refusal of peasant women to remove the filth from their children's heads,[19] without always realizing that such practices owed their credibility more to recent medical knowledge than to the medieval Salernitan precepts. The Montpellier doctors condemned the harmful effects of thoughtless use of water. Overfrequent ablutions and, a fortiori, baths weakened animalization and therefore sexual desire. Bordeu cited vigorous, "odoriferous" individuals whose "skin was cleansed, their strong emanations and transpiration destroyed; everything that distinguished their sex was deadened."[20] Moreover, the *aura seminalis,* and therefore the power of attraction, "is better preserved in unkempt individuals who do not waste their time and sap by cleansing themselves." Bordeu warned city dwellers against "the luxury of cleanliness," particularly pernicious to women in childbirth and to invalids with "sweating" diseases.[21]

In reviving these principles, Brieude set the seal of his approval on a traditional behavior pattern that some theorists at the time were striving to change with arguments derived from notions of the delicacy of the senses[22] or the need for the deodorization of public space.

All the vapors and emanations expelled by the excretory ducts made up the atmosphere of the individual; doctors would reexamine the most aberrant examples for two centuries. Socrates had noted that young brides did not need to use perfumes, since the sweetest scents emanated from their person. Alexander's body smelled of violets, Montaigne reported on Plutarch's authority;[23] Haller exhaled musk; M. de la Peyronnie, according to the *Encyclopédie,* "knew a man of rank whose left armpit exuded a surprising odor of musk during the summer heat."[24]

Individual atmosphere was seen to vary according to a series of factors linked to an anthropological theory whose concepts and mechanisms are wholly familiar today;[25] but this anthropology was deeply rooted in contemporary medical theory. Variations in the smells of living creatures resulted from the composition of the humors, the functioning of the organs, and the strength of the powers of depurgation. Anything that influenced one of these elements caused a change in the odor emitted by the individual. "The clime he inhabits, the seasons he encounters, the food he eats, the passions he indulges in, the type of work he pursues, the arts he practices, the earth he digs, finally, the air he breathes, change in different ways the humors he assimilates as well as those he exhales; which necessarily results in different odors."[26] This anthropological theory did not postulate the radical inferiority of certain races but at most their "degeneration"; altering one of the variables was enough to change the odor of the body.[27]

Human beings possessed a succession of smells from childhood to old age, from the milky sourness of the suckling to the sweeter, less acid, sourness of senility, a smell that Haller found intolerable.[28] Between the two extremes was the fragrance of adolescence, particularly marked in young girls. Puberty, which radically transformed the odor of males and gave them the *aura seminalis* of the adult, did not change women's constant odor so clearly. "At that time their slack, rarely exercised fiber only dulls their childhood sourness and gives a stale and sweetish odor to their transpiration."[29] Nevertheless, menstruation and, above all, sexual intercourse, temporarily altered the character of their smell.

Oddly, medical discourse concerned itself little with the specific relation of smell to temperament, the color of hair, or complexion. The odor peculiar to irascible personalities and the smell of redheads were noted,[30] but without emphasis, as if they were self-evident. Passions exercised an effect on the humors; they also affected individual odor. Some passions operated slowly but profoundly; they checked organic movements and cut off secretions. Thus people lost their odor when they were sad. Passions that struck by fits and starts intensified the bodily stenches. Accelerated putrefaction of the bile reinforced the smell of breath when a person was angry. Terror gave underarm sweat a foul smell and created intolerable wind and stools. Classification of the glutton's fetidity, the drunkard's vinous sourness reinforced the traditional view of the stench of the sinner that had enabled Saint Philip Neri to recognize souls destined for hell. Thus the belief in the fragrance of the saints was supported *a contrario*.[31]

The ingesta, that is, air, food, and drink, regulated the excreta and therefore individual odor. "The Negro and the Samoyed, like the filthy Hottentot, *must* to some extent smell more strongly."[32] They represented the brutish, strongly animalized world. "In the torrid zone, the Negroes' sweat always has such an evil-smelling odor that it is hard to stay near them for a few minutes. The Finns and Eskimos who live near the poles spread an intolerable stench around themselves."[33] The same was true of Cossacks.[34] Virey was more specific: "Negroes from certain West African regions, such as the Jolofs, exhale a stench of leeks when they become too hot."[35] When blacks and whites bathed together in the tropics, the former were more exposed to rapacious sharks because of the odor they emitted.[36]

Common to all these comments were the references to climate, quality of air, and degree of putridity of foodstuffs—in short, the mechanism of depurgation.[37] The Samoyeds were as fetid as the blacks, not because the climate in which they lived accelerated decomposition of the humors, but because these people were fond of putrid foods.

Such analyses differed fundamentally from anthropological discourse at the end of the nineteenth century. Witness the fact that identical observations were made within France independent of any reference to racial origin, state of poverty, or poor hygiene. If the odor given off by a band of Cossacks could be discerned hours after it had passed, so could the effluvia from troops of "cowherds from our mountains."[38] The odor of rural people differed from that of city dwellers; the humors of the former were less corrupt, "closer to vegetable nature."[39] People who liked meat smelled foulest of all; for the moment, these tended to be town dwellers.

Regional populations exhaled specific odors, again as a result of the kind of food they ate. "When harvest time brings these people together in our cantons, it is easy to tell men from the Quercy and Rouergue regions by the fetid odor of garlic and onion they give out, while the odor of the Auvergnats is like soured whey."[40] On the whole, these odors were more pronounced in southern regions.

People's occupations also gave rise to a gamut of individual odors mediated by the substances handled at work. Peasants could be identified by their scent, noted Brieude; and everyone, he asserted, knew the sweetish odor of nuns' cells, symptomatic of a "weak or imperfect assimilation." "Who," he asked, "could not tell a cesspool clearer, tanner, candlemaker, butcher, etc., solely by the sense of smell . . . ? A certain quantity of those volatile particles which penetrate the workers is expelled from their bodies almost intact, along with their humors, with which they probably partly combine . . . The odor that results is the very sign of the health of these workers."[41] Brieude diagnosed disease among several tanners solely from the fact that they had lost the odor of their trade. This did not lead him to feel any social repugnance. He simply integrated the results of his observations into the picture of the different occupations that sanitary reformers had been striving to produce since Ramazzini.

The odor of bodies also became an element of medical semiology. Hippocrates had already classed it as a symptom.[42] Invasion by disease could be diagnosed both by the loss of a healthy odor and by the appearance of a morbid one. The progression to disease and then to death went from the acidic to the alkaline condition of the putrid matter.[43] While deploring the dearth of an appropriate vocabulary to define the odors perceived, Bordeu stated that the medicine of his day "assessed the essence of the parts and their healthy or diseased state by the sense of smell."[44]

The doctor was therefore not content to be an expert in mephitism; when he went to his patient's bedside, he still had to learn to "smell reflectively."[45] Before he did anything else, he had to carry out a difficult calculation: he had to establish how the patient *ought* to smell, taking account of age, sex, temperament, hair color, occupation, and, if possible, individual odor, recorded when the patient was in good health. After that the practitioner referred to the progression of smells that characterized each type of disease. The patient's odor then enabled him to make his diagnosis and prognosis. Analysis of smells obviously favored "issues," notably breath, stools, and, above all, pus, which proved surprisingly amenable to interpretation. "One notices every day when dressing wounds, and even all cutaneous suppurations, that the properties of the suppuration change if an invalid has given way to violent passions, if he has taken too heavy or too lengthy exercise, if he follows a poor diet, if, above all, he indulges too freely in strong liquors, if he lives off sharp, salted, or smoked foods, if he inhabits infected or swampy air."[46]

The masses had long monitored themselves for smells as symptoms of disease. Midwives and domestic servants, particularly in the countryside, spontaneously reported to the doctor changes in odors of sweat, stools, urine, sputum, ulcers, or linen that had been in contact with invalids' bodies.

All the authors who compiled interminable catalogs of smells condemned the stench from scurvy as the worst. Moreover, "qualified practitioners can very well tell the odor that emanates from ulcers complicated by gangrene, every odor peculiar to consumptives, people laid low by dysentery, malign putrid fevers; and that odor of mice which is part of hospital and jail fevers."[47] When the sour, milky odor of the woman in childbirth turned fetid, a milk fever could be diagnosed.

Analysis of the contents of these works or even a simple lexicological count would probably make it possible to understand the field of osphresiology better and to measure the enlargement of its vocabulary, which seems to have reached its peak during the Restoration. After the publication of Pringle's work, there emerges a visibly coherent set of relations between the practices employed by a large number of doctors, the theoretical study of putrefaction, the anthropological discourse applied to osphresiology, and a type of spontaneous therapeutics in use among the masses. Humoral medicine, already a thing of the past, the theory of putrid fevers, the vitalism of the Montpellier school, and Bordeu's organicism were a long way from being unanimously accepted in medical circles, but they reflected very clearly the deep-rooted belief in the importance of individual odors.

Next it was the turn of the chemists to attempt to analyze this fascinating aura. Before Lavoisier and Séguin had published their analyses of the respiratory exchanges that took place through the skin, there were numerous often confused, usually amusing, attempts by others to solve two related problems. Two centuries earlier, Sanctorius had proved the existence of insensible perspiration by calculating the loss of its weight upon a body; was it possible to reveal it by measuring and analyzing the gases evacuated through the skin?

Could the inhalation of odoriferous vapors be proved in the same way? The answers, it was falsely hoped, would throw light on the mechanisms of infection and contagion.

Accordingly, chemists, including the most eminent, decided to immerse themselves in warm water in their baths, their bodies enclosed in glass jars, in order to collect the gases from their arms, armpits, or intestines. In 1777 the comte de Milly presented the Berlin Academy with an analysis of the gases that had emanated from his skin; he claimed that it was "fixed air." Cruickshank and Priestley emulated him, although the latter had little confidence in this empirical approach. In 1780 in Paris, and then in Baden, Ingenhousz collected the gas that seeped through the skin on his arms; he thought he recognized in it Priestley's "phlogisticated air." He plunged a young girl of nineteen into a bath and decided that the emanations were no less noxious than the air he had collected from his armpits.[48] Thus the assumed therapeutic quality of "young air" was disproved. Jurine set out to refine the analysis. He carried out repeated experiments on children aged ten to nineteen, men aged thirty-six to sixty-five, and one woman aged forty. In each case he collected a gas that he called aerial acid, whose function, he claimed, was to rid the body of its phlogiston.[49]

Numerous aberrant attempts devoid of scientific interest nevertheless provide evidence of the scientists' passionate quest to give a scientific basis to their beliefs. In Italy, Canon Gattoni, a student of Volta, set out to measure the deterioration in the air caused by infirm and unhealthy bodies.

> I took a few young beggar boys and, by employing a greater or lesser amount of money, I was able to induce them to let themselves be enclosed in large leather sacks, a sort of leather bottle, up to their loins. I then had the sacks squeezed as tightly as possible around their bodies. The better to cut off communications between internal and external air I had sheets soaked in water sewn to the openings of the sacks, and I left the young prisoners in this uncomfortable posture as long as they could stay. After that I made them plunge up to their stomachs in a warm bath in a tank under a large funnel already prepared to receive the air confined in the sacks (which were collected in large glass vessels), methodically decanting the air thus created for analysis by the eudiometer.[50]

Both Jurine and Gattoni sought to prove through their analyses that foul-smelling intestinal emanations were noxious, and they used

nearly identical methods to collect the gases. Jurine analyzed those
that remained in the bowels of corpses. Their results proved the existence of cutaneous respiratory exchanges.[51] For the time being this
finding could only strengthen belief in the inhalation of miasmas;
but chemists using the eudiometer did not succeed in accounting for
variations in individual odors. Bordeu triumphantly scoffed at the
fashionable study of breaking wind: the doctor's nose proved superior
to the scientist's apparatus.[52]

Nonetheless, one result seemed unquestionable: stenches could
be inhaled by the organism and affect individual odor. To us this
seems a derisory conclusion, but at the time it acquired intense
emotive power. Bichat bore witness to the phenomenon:

> I observed that at the end of a period in the dissecting rooms, my
> wind frequently took on an odor identical with the odor exhaled
> by decomposing corpses. Now, I ascertained that it was the skin as
> much as the lungs that absorbed the smelling molecules at that time,
> in the following way: I held my nostrils and fixed a fairly long pipe
> that went from my mouth out of the window, to enable me to
> breathe outside air. And what did I find? After I had remained one
> hour in a small dissecting theater, next to two very fetid corpses,
> my wind produced an odor very similar to theirs.[53]

This intensive olfactory vigilance regarding personal emanations
confirmed belief in the mysterious penetration of the body by putrid
effluvia. Thereafter anxiety about other people's odors became pre-
dictable.

The Economy of Desire and Repulsion

The atmosphere of bodies influenced human
relationships at two levels: at the level of personal attraction or revulsion, and at the level
of infection. As early as 1733, Philippe Hecquet, dean of the Paris Faculty of Medicine,
explained the convulsions of the patients at Saint-Médard as a result
of erotic stimulation arising from the impact of corpuscular emanations from the convulsionaries. In 1744 Hartley attributed sexual
desire to vibrations that acted on the neural fibers. On the eve of
the Revolution, the arousal of attraction or repulsion through personal odor was still a literary theme, as attested by the sympathists'
theory.[54] Tiphaigne de la Roche held "that particles of an invisible
substance called sympathetic matter spread around men and women;

that these particles act on our senses, and that this action produces attraction or aversion, sympathy or antipathy, with the result that when the sympathetic matter which spreads around a woman, for example, makes a pleasing impression on the man's senses, thenceforth that woman is loved by that man." According to Tiphaigne, this sympathetic matter was none other than the doctors' "transpirant matter." Every individual was thus linked to the rest by an infinite number of threads that "titillated" or "tore" their fibers.[55]

An idea inspired by the same sympathist theory can be found in a fable in Mirabeau's *Erotika Biblion*. In it, every inhabitant of the rings of Saturn exhaled his own specific effluvia, which were directly linked with the "nervous buds of perception." These emanations could intertwine in other people's effluvia, producing "live cohesion" of two beings by innumerable similar molecules. In the rings of Saturn, both knowledge and feelings were transmitted through the air.[56]

That odor governed attraction was a commonplace. Casanova confessed to its fascination.[57] The sense of smell became the sense of amorous anticipation, of vague desire that could prove false when sight supplied the details; an example was Don Juan's being misled by Elvira's *odor di femmina*.[58] In 1802 Cabanis christened the sense of smell, the sense of sympathy; at the end of the nineteenth century Monin and Dr. Galopin still considered it the sense of affinities.[59]

Once again, experts had set their seal of approval on ancient stereotypes. The male's *aura seminalis* stirred female desire, just as it kept man's appetite alive. It was by "the perfume of incense and rose mixed" with "fresh living blood" that Paris drove the palace ladies mad with hunger for his breath.[60] The status of male desire was actually more complex than this. The model of the rutting animal was an obsession; doctors remained convinced that seduction owed a great deal to the odor of the menses. The ambivalence toward menstruation is a familiar theme. Menses were an aspect of the purging process and therefore exercised a putrid effect; but they were also impregnated with subtle vapors transmitted by the essence of life.[61] From the viewpoint of the Montpellier school the woman at that point in her cycle was conveying the vitality of nature; she was emitting the products of a strong animality; she was making an appeal for fertilization, dispersing seductive effluvia. Decades later, in already outdated terms, Cadet de Vaux praised woman's atmosphere as "the guiding spirit exuded by the essence of life contained in her reservoirs."[62] The special condition of women whose smells were in a state of imbalance also derived from these beliefs: redheads were always pungent, both putrid and fascinating, as if their cycle had broken down and put them in a continuous state of menstruation; pregnant women, temporarily deprived of menstrual effluvia, were also examples of imbalance.[63]

This belief in the vitality of menstruating women was later shared by Michelet, who was fascinated by the appearance and flow of his young wife's periods. But in the meantime an upheaval had taken place. Spontaneous ovulation, suspected as early as 1828 and confirmed by Pouchet in 1847, defused fear of the sorceress whose menstrual effluvia tarnished metal and turned the meat in the salting-tub. This woman was no longer evil; she too had become a creator. Menstrual blood was thenceforth charged with new significance. Husbands, like gynecologists, chronicled the cycle of blood and a mysterious life hitherto perceived only by women. Woman had gained a new innocence from having her periods and odors glorified by scientific discourse, but she had lost her secret powers.[64]

When menstruation was not occurring, the vital vapors that impregnated menstrual blood perfumed the other emunctories. These effluvia pervaded the poetic discourse of the period. Odoriferous tresses enchanted Parny and Bernis well before Baudelaire.[65] The seductive power of underarm sweat and of an impregnated chemise inspired innumerable anecdotes. Henri III, it was said, remained in love with Mary of Cleves after breathing the odor of her linen in a closet where she had just changed; the olfactory message engendered the amorous *coup de foudre* with a suddenness comparable to the effect of the abruptly torn curtain described by Roland Barthes in connection with the phenomenon of romantic love. The prince fell in love with the perfume of flesh, as did Werther with the tableau of Charlotte framed in the door.[66] One oriental sultan chose his favorites by the perfume of their sweat-soaked tunics.[67] Goethe confessed to stealing one of Madame von Stein's bodices so that he could breathe it at leisure.[68] Barbey d'Aurevilly's "Bewitched" tried to seduce her abbé by sending him one of her chemises. Huysmans *(Le Gousset)* cried out his fascination with the odor of female armpits.[69] More subtle was the seductiveness of the bouquet laid on the mistress's breast; both Rousseau and Parny sang hymns to its ravishing effects.[70] And the odor of shoes delighted Restif de La Bretonne well before experts had invented leather fetishism.[71] But there is one strange silence, probably reflecting a taboo: these erotic writings make no allusion to the seductive power of vaginal odors except for a few references to menstruation.[72]

Puberty was not the crucial stage in the lifelong progression of a woman's bodily odors. As Yvonne Verdier has noted, menstruation increased the young adolescent girl's seductive power and was a reminder of her creative mission, but it endowed her with only a discontinuous odor. What gave woman's smell its ideal quality was male sperm, in the same way that the act of coitus filled the flesh of the females of numerous animal species with a specific odor.[73] In every sphere, sexual intercourse was held to complete femininity.[74]

A few endlessly repeated anecdotes about individuals' exceptional olfactory discernment are revealing. They concern not menstruation—which perhaps revealed itself too obviously—but sexual activity, spermatic impregnation of a woman's organs and humors, and therefore the excretion of seminal vapors. After a report in the *Journal des Savants* in 1684, all medical works on this subject repeated the example of the monk of Prague who was capable of spotting the odor of adulterous women.

Excessive indulgence in coitus provoked a positive overflow of sperm into the woman's humors, putrefied the liquids, and engendered an intolerable stench. That was how prostitutes became *putains*.[75] Juvenal had already made this claim; at the beginning of the eighteenth century J.-B. Silva tried to find a scientific justification for this belief, which in itself was enough to cause prostitutes to be considered dangerous women.[76]

Sick bodies and live bodies that were rotting, especially those of animals, did in fact emit morbific odors. When epizootic was prevalent, man too was in danger. Far from being beneficial, the air of stables came to be regarded as very unhealthy.[77] The theory of putrid fevers ran counter to vitalist convictions.

Proximity to a sick person was even more dangerous. Around the invalid there formed an atmosphere "varying in extent, which clings to his clothing, his furniture, the walls of his room; which is heavy, oppressive, less mobile and elastic than ordinary air, and which lingers in the corners of apartments for a very long time." The stench was a danger signal enough. During the epidemic of camp fever that ravaged the French army billeted at Nice in 1799, the "unfortunate soldiers emitted an odor similar to that of phosphorus gas in combustion, which could be smelled from a long way off, and which

lingered in the streets and houses where most of the patients were."
Convalescents themselves propagated the disease, Fodéré stated, insofar as they were not "stripped of their atmosphere."[78]

But it was breath more than anything else that conveyed miasmas
and stenches.[79] Boissier de Sauvages advised distrust of the fetid
breath of "the diseased animal." The breath of a moribund cesspool
clearer struck down one of Jean-Noël Hallé's companions.[80] Breath
showed the presence of life and its seductions. Correct respiration
ensured the influx of "vital air"; but it also became the emunctory
for the dirt accumulated by the humors, the outlet for "phlogisticated
air" that vitiated the environment. In those days of obsession with
mephitism and miasma, other people's effluvia, breath, and body
odor were subject to particular attention.

To contemplate the mass of vapors that accumulated where living
beings crowded together was to be seized with a vertiginous sense
of alarm. Perception of the danger from "social emanations" brought
distrust of the putrid crowd, of people and animals combined.[81] Calculations of these emanations supplanted the weighing of terrestrial
vapors. As early as 1742 Arbuthnot tried to measure city dwellers'
exhalations: "the sweat of less than 3000 Human Creatures would
make an atmosphere 71 feet high, over an Acre of Ground, in 34
Days. This perspirable Fluid is to Air in Density, perhaps as 800 to
1; therefore if you extend the 3000 people over a hundred Acres of
Ground, there will remain 8 inches, the greatest Part of it remaining
unblown off, and spread with the infinite Tenuity of odourous Effluvia, will infect the whole Air of that City."[82] Boissier de Sauvages
resumed the calculation nine years later: "reduced to vapors," the
excrement produced by the five pounds of food the city dweller
ingested daily formed a column (which weighed five pounds) four
feet seven inches high around his fifteen feet of skin. In the towns
this column was twice as dense.[83] In fact the area available to every
inhabitant was no more than half the area of his epidermis.

The mere presence of bodies, even of healthy bodies, both vitiated the air in towns and infected the air in valleys. Fortunately the
wind, the cleansing effect associated with the movement of carriages,
and the combustion in hearths stirred up the air and partly counteracted the density of the atmosphere. This was not the case in closed
spaces teeming with crowds. Epidemics that wreaked havoc in towns
originated on ships and in hospitals, prisons, barracks, churches, and
theaters. This obsession with the crowding together of bodies later

governed both urban social perceptions and the tactics that sanitary reformers used haphazardly in public space until Lavoisian chemistry allowed more precise norms to be determined.

The Odors of the Sick Town

The image of the nauseous bilge obsessively informed social perceptions of putridity. "The complex odor so characteristic in ships is a product of emanations from the hold, the aroma of tar, the general fetidity engendered by so many men collected together in a small space."[84] Thus the sanitary reformers applied their hesitant analyses to the case of ships, where, Hales claimed in 1743, "the bad Air," like the air in jails, often caused "infectious pestilential distempers."[85] A few years later the vicomte de Morogues tried to analyze the complex odors that, in his opinion, were sufficient to explain the devastating putridity of scurvy.

The ship was a "floating swamp."[86] Seawater seeped through the seams in the bulwarks; fresh water stagnated in puddles after rainfall or ill-considered washing, soaked the rigging, eroded the timbers, oxydized the cannonballs, and combined with the ballast to form a blackish and murderous ooze. Or else the vile-smelling liquids built up in the bilge, which synthesized every stench. The fetid odor of this mixture of fresh water and seawater was exacerbated when the pump was working; it was as noxious as the smell of deserted salt marshes. Sea bitumen, mists from the shores, and the exhalations from harmful mooring grounds promoted the identification between ship and marsh.

It was also a choice area for fermentation. Vapors were exhaled by the wood in the frame and the hemp in the rigging, especially when the ship was new. In the hold where the victuals were kept, "there is always the smell of a warm vapor and an unpleasant odor that could make a delicate person faint."[87] This was a mistaken sensation: the air in the hold only appeared to be hotter. Stagnation aggravated the infection. The bottom of the hold was a constant threat to the life of the vessel. It exercised a strange fascination. Much later, Arthur Gordon Pym's embarkation, Dracula's crossing, and the foul cargo in Conrad's *The Shadow Line* attested to the horror of these malodorous depths. "Emanations produced by strongly smelling or fermented liquids and victuals" also escaped from the hold.[88]

The dung and sweat of the animals on board, the poultry droppings, the supplies of fish, the decomposing corpses of rats and insects, the rubbish that had slid under chests and piled up in dark
corners generated a mélange of foul odors that threatened sailors
and passengers.[89] In hot weather the smells from the "bottles" and
pots of urine became intolerable. *Arthur*, a ship carrying poudrette
(dried excrement) from Guadeloupe, lost half its crew to its evil-
smelling cargo in 1821. It reached the open sea at Pointe-à-Pitre like
a ghost ship.[90]

Crowded bodies and the vapors from combustion completed this
potpourri. At night the sailors squeezed together on the orlop deck;
they slept in stagnant air in clothing impregnated with humidity and
sweat.[91] So oppressive was the odor that passengers practically suffocated when they passed a hatchway. The unhealthy stench of the
ship was held to resemble the stench of the infectious quarters of
towns. The ship's infirmary was sometimes transformed into a floating
hospital. Sailors in irons moldered in its holds. When the cargo
consisted of blacks from Guinea, Hales emphasized, the air was
"intolerably nauseous . . . especially when the Ports are shut."[92]

Scientists set out to weigh the vapors exhaled on ships.[93] They
endeavored to define the complex smells produced by this powerful
amalgam and to measure its dangers. In 1784 a commission from the
Société Royale de Médecine studied the problem.[94] As a result of
its recommendations, foul-smelling vessels were burned even if they
were seaworthy. Such was the fate of the frigate *Melpomène*.

On land the worst olfactory scandal was the prison. The stench
was the sign of the living and collective putrefaction of the convicts.
The human muckheap combined both a genealogy of past infection
and a continuing putridity in the present. According to Louis-Sébastien Mercier, Bicêtre could be smelled four hundred *toises* away.[95]
One of Cartouche's accomplices feigned death in order to be freed
and enabled to breathe fresh air for a few seconds. The comte de
Struensee, taken out of his dungeon to be beheaded, cried out: "Oh,
the happiness of breathing fresh air!"[96] In the terrible Plombi jail in
Venice, the disputes between Casanova and his jailer centered on
removing the stenches of his bucket.[97]

Prisoners' access to pure air was a major issue at the beginning
of the nineteenth century. Senancour's Obermann compared himself
to someone who, after ten years' detention, left the vileness of the
dungeons and saw the serenity of the sky again.[98] Pizarro's prisoners

sang their joy at coming out into the air and light.[99] Michelet alluded to the history-laden odors of incarceration on several occasions: "The old, humid, dark convents that are used for this purpose almost everywhere today retain an indestructible essence of historic uncleanliness, an indefinable odor that sickens the heart from the moment of entering, whatever one does. The poor wretches who experienced Louis XIV's prisons said that their vitiated air was the greatest torture."[100]

In 1784 Howard complained that "methods are continued to rob prisoners of this *genuine cordial of life,* as Dr. *Hales* very properly calls it . . . The air of the prisons infects the clothing of those who visit them . . . A vial of vinegar has, after [being used to ward off infection] in a few prisons, become intolerably disagreeable."[101] Consequently, this greatest of the prison observers was constantly vigilant for smells.

The scandal attached to the stench of incarceration was already of long standing when Howard wrote. Bacon considered the smell of a jail to be the most dangerous infection after the plague.[102] The long list of catastrophes was abundant proof of this claim. At the Black Assizes at Oxford in 1577, Roland Jenkins was sentenced for seditious talk. From the courtroom "such a pernicious vapor arose that all who were present died within forty hours."[103] Taunton assizes proved equally lethal in March 1730. Prisoners brought from Ivelchester infected the court, and the chief justice, the solicitor, the sheriff, "and some hundreds besides, died of the gaol-distemper."[104]

The assizes held at the Old Bailey on May 11, 1750, were among the most terrible. Before the hearing, the prisoners—there were two hundred of them to follow each other in—had been thoughtlessly crowded into two rooms that gave on to the judge's chamber and the "bail dock," a sort of alcove connected to the court by a door and by an opening at the top of the partition. These three rooms "had not been cleaned for some years. The poisonous quality of the air still aggravated by the heat and closeness of the court, and by the perspirable matter of a great number of all sorts of people . . . the bench consisted of six persons, whereof four died, together with two or three of the counsel, one of the under-sheriffs . . . and others present, to the amount of forty in the whole," not including "those of a lower rank, whose death may not have been heard of."[105] Again in 1812, the assizes at Lons-le-Saulnier turned into tragedy.[106] Both Pringle and Lind thought that the infections that ravaged his majesty's fleet and armies sprang from prisons.

Before Priestley's discoveries, the difference between that terrible "jail fever," produced in stenches, and pure and simple suffocation due to overcrowding was ill defined. Scientists confused the tragic destiny of the 117 English prisoners suffocated in Calcutta, in the "black hole" in which they had been shut up,[107] and the diseases that overwhelmed the victims of the deadly assizes.

Prison odors and the fever they engendered were all the more formidable in that they resulted in part from past impregnation. In these circumstances the only solution was to abandon the sites and demolish the buildings. That was the conclusion of Lavoisier's report to the Académie Royale des Sciences, after his visit to Saint-Martin and For-l'Evêque.[108]

Half a century later, the discourse organized around the putridity and stench of the prisoners' cells provided the pattern for descriptions of urban workers' dwellings and peasants' ill-kept houses. The dungeon formed the model in terms of which the interminable but justified diatribe against insalubrious living conditions was formulated as early as the eighteenth century.

According to contemporary observers, what characterized the smell of the atmosphere of the hospital was the complex nature of the putrid odors.[109] Patients' quickened respiration and foul-smelling sweat, their purulent sputum, the variety of pus that flowed from wounds, the contents of buckets and commodes, the pungency of medication, the effluvia of plasters, all amalgamated into a stench that the practitioner tried to analyze as fast as possible in order to avert the risk of an epidemic. The sex, age, occupation, and temperament of the patients modified this overall fetidity, from which the effluvia of the dominant disease emerged. The worst was really "hospital fever," the odor of corpses that preceded and foreshadowed death; it rose from gangrenous limbs and from the sweat-impregnated beds reserved for the dying.[110]

Visiting and describing hospitals became initiation tests for everyone involved in public health; and the suffocating odor was given pride of place. The indefatigable Howard, insensitive to noise, indifferent to lighting, continued to analyze smells, as in his descriptions of the hospitals at Lyons and Malta.[111] Madame Necker was not afraid to visit that part of Bicêtre reserved for the sick, and notably "Saint-François's ward," a veritable "Parisian type of bilge" where the stinking air "made the most charitable and most intrepid visitor faint and suffocate."[112] All hospitals, asserted Louis-Sébastien Mercier, were evil-smelling. This testimony was confirmed by the

ran the risk of disease. Horsehair workers had chosen an unhealthy trade. Fullers, who used excrement, "are continuously in very hot workshops, surrounded by the evil-smelling odors of decayed urine and oil, often half naked; almost all of them become cachetic";[128] the putrid molecules spoiled the density of their blood. The worst—already—were those scavengers who traded in animal remains.

But the majority of laborers were healthy and worked in odorless workshops, far from fermentation and putrefaction. The fundamental factors in Ramazzini's classification, completed by Fourcroy, revived by Patissier and, partially, by Parent-Duchâtelet, were still the nature of the product handled, the quality of the surrounding air, and the composition of the vapors inhaled. Health depended on these, as it depended on the worker's food, climate, or temperament. It was never attributed to poverty, specific living conditions, or membership in a social class; nor was it attributed to a race with a predetermined biological destiny.[129] Being a worker did not automatically imply stench; it was up to the worker, like anyone else, to protect himself from fetid surroundings.

At the beginning of the century, Le Sage in *Gil Blas* was already condemning the odors of Madrid, in reality those of Paris, but his account never assumed the tragic aspect of Louis-Sébastien Mercier's:

> If I am asked how anyone can stay in this filthy haunt of all the vices and all the diseases piled one on top of the other, amid an air poisoned by a thousand putrid vapors, among butchers' shops, cemeteries, hospitals, drains, streams of urine, heaps of excrement, dyers', tanners', curriers' stalls; in the midst of continual smoke from that unbelievable quantity of wood, and the vapor from all that coal; in the midst of the arsenic, bituminous, and sulfurous parts that are ceaselessly exhaled by workshops where copper and metal are wrought: if I am asked how anyone lives in this abyss, where the heavy, fetid air is so thick that it can be seen, and its atmosphere smelled, for three leagues around; air that cannot circulate, but only whirls around within this labyrinth of houses: finally, how man can willingly crawl into these prisons whereas he would see that, if he released the animals that he has bent to his yoke, their purely instinctive reaction would be to escape precipitously to the fields in search of air, greenness, a free soil perfumed by the scent of flowers: I would reply that familiarity accustoms the Parisians to humid fogs, maleficent vapors, and foul-smelling ooze.[130]

Completing this list were prisons, churches, the fetid cesspools along the Seine (on the quai de Gesvres, for example), and, above all, the markets, miscellanies of smells in the heart of Paris. From 1750 on, Les Halles became one of the favorite sites of the new vigilance.[131] The underground storerooms exuded a complex range of odors from rotting vegetation. At ground level, in the "porte merdeuse" sector, the effluvia of fish assailed passersby. The impregnation of the stalls by foul smells aroused the—largely phantasmic—desire to destroy that is so frequently manifested in the projects of urban reformers.

Observers attempting to analyze the hitherto impenetrable stench of the center of the capital produced many unexpectedly precise accounts. The cumulative result is an incomplete, discontinuous olfactory perspective, severed from the dominant harmonizing logic of the visual.[132] By mapping the flux of smells that made up the olfactory texture of the city, these observers located the networks of miasmas through which epidemics infiltrated the capital. Much later, this new view of urban space gave rise to a fresh reading of society. But at the time the sociological project remained somewhat indeterminate. The urgency of the dangers revealed by the confused mixtures of odors from earth, water, excrement, corpses, and living bodies hampered analysis. Not until the nineteenth century did sanitary reformers use tactics that created a clear distinction between the deodorized bourgeoisie and the foul-smelling masses.

On the eve of the Revolution the essential aims were different: since the aggregation of miasmas increased the danger tenfold, the confusion of smells had as far as possible to be destroyed; the task was to categorize the threats. The architect Boffrand envisaged building as many markets as there were products.[133] Chemists set out to analyze the air in crowded places, hoping to prove that Pringle's theories were well founded. For a long time the process remained clumsy and unproductive. Even in Priestley's hands the eudiometer proved incapable of measuring the degradation of the air in workshops or ships' holds. Volta and Gattoni were slightly luckier; basing their work on the speed of combustion, they determined the quality of air in the sickroom. After having laboriously attempted to analyze the air from sick beds, Jurine used the same method to draw up a scale of putrid places in general. At the top were dungeons, henceforth scientifically considered the most dangerous.[134] But scientists already understood that all these measurements served to determine only the respective quantities of "vital air," "inflammable air," and "chalky acid." Miasma remained elusive.

But the new multiplicity of references to smell in scientific discourses does not mean that Louis XVI's subjects were assailed by more intrusive odors than in the past. The only firm conclusions we can draw are of a phenomenological sort, based on an account of subjective perceptions. From about the middle of the eighteenth century, odors simply began to be more keenly smelled. It was as if thresholds of tolerance had been abruptly lowered; and that happened well before industrial pollution accumulated in urban space. All the evidence suggests that scientific theory played a crucial role in this lowering of thresholds. We have lost sight of that connection, because we operate with a history of science that favors the discovery of scientific truth and neglects the history of scientific error.

In measuring the scale of the new sensibility, however, there remains one area of uncertainty. Everything in the preceding pages simply proves a new alertness to the olfactory environment within a very specific milieu: that formed by doctors, chemists, and reformist campaigners. To be sure, this sample is not unique; it points to a socially more widespread change of attitudes. Nevertheless, the exact extent of the social diffusion of this new attitude of anxiety and vigilance does need to be more precisely charted.

4 Redefining the Intolerable

Lowering the Thresholds of Tolerance

ONE BASIC FACT needs to be considered from the outset: the warnings deployed by the experts to expose the risks of infection had immense force. The ideal of pure air made the peril seem urgent and revived (and sustained) the specter of urban suffocation. "The need is pressing," clamored Tournon; "the capital is nothing more than a vast cesspool, the air is putrid there . . . In some [districts] it is already so foul that the inhabitants can hardly breathe."[1]

The campaign formed part of a social education in hygiene, which emerged at the same time as the leap forward in chemistry (1760–1769); Daniel Roche has meticulously traced its progress. Inspired by a "mystique of utility"[2]—it is too early to talk about utilitarianism—experts of all kinds observed, collected, kept records. They embarked on a vast inventory of smells that—along with the other administrative inventories of the period—sprang from the desire to ensure the efficient functioning of the city; thus health administration was built upon a catalog of noxious odors.

But the influence of the anxious outcry from enlightened elites must not be overestimated. We must not overlook the indifference of the masses to the issue of smell, and their resistance to the tactics of deodorization (discussed more fully later). Observers who were surprised at the tolerance of stench attributed it to familiarity—proof that behavior patterns were shifting. "There is no one in the world like the Parisian for eating what revolts the sense of smell," exclaimed Louis-Sébastien Mercier, scandalized by the fish shops in the capital.[3] No unpleasant odor could repel the Paris tradesman, noted Chauvet, "so accustomed is he to infection."[4] Young girls strolled and chatted pursued by the exhalations of the corpses piled up in the cemetery of the Innocents; "they can be seen buying fashions, ribbons . . . amid the fetid and cadaverous odor that offends the sense of smell."[5] Little girls in the parish of St.-Eustache heard the catechism without being repelled by the noxious emanations.[6] The report written by the Paris priests to oppose removal of the dead bore the imprint of this relative mass anesthesia.[7] Nonetheless the fact remains that tolerance of this "frightening proximity" was henceforth labeled "strange."[8]

Arthur Young's account conveys a Briton's amazement at the olfactory tolerance shown by most Continentals. "[On the contrary,] go in England . . . your senses may not be gratified, but they will not be offended."[9] At the inn of Pezenas, he wrote, "we were waited on by a female without shoes or stockings, exquisitely ugly, and diffusing odours not of roses: there were, however, a Croix de Saint-Louis and two or three mercantile-looking people that prated with her very familiarly."[10] Stranger still in his eyes was the attitude of the Clermontois: "The contention of nauseous savours, with which the air is impregnated, when brisk mountain gales do not ventilate these excrementitious lanes, made me envy the nerves of the good people, who, for what I know, may be happy in them."[11]

Yet the end of the eighteenth century also brought the first symptoms of a reduced threshold of tolerance among the masses, who made a direct connection between odors and death. The long-established behavior patterns displayed at times of plague are clear proof of this attitude.[12] Menuret had noted in 1781 that "the crowd hastens to avoid the odor and venom of disease and of death."[13] The stench of corpses seems to have been the first to arouse nearly universal intolerance, as the early date of complaints by people living next to cemeteries attests. The perceived correlation between the odor of corpses and the corruption of meat and metals accentuated feelings of anxiety and explains the vehemence of protest with which those feelings were expressed. To separate the abode of the dead from that of the living became an incessant demand. Today this movement is well established as a significant episode in the history of public opinion. Madeleine Foisil has described the complaints provoked in 1672 by exhalations from the cemetery of the Trinité.[14] Philippe Ariès, Pierre Chaunu, and all the other experts on attitudes to death in former times have emphasized the force of the campaign to keep corpses at a distance, especially in relation to Paris. Petitions from people living next to cemeteries back up scientific treatises and administrative reports. It was a series of complaints from the masses orchestrated by the stallholders in the rue de la Lingerie that finally brought about the closing of the cemetery of the Innocents in 1780.[15]

The lowering of the threshold of olfactory tolerance is a well-perceived and well-described fact. Louis-Sébastien Mercier analyzed its mechanisms lucidly though with some contradictions; he attributed it to the chemists. "Twenty years ago one drank water without paying much attention to it; but since the family of gases, the race of acids and salts have appeared on the horizon . . . one has armed oneself on all sides against mephitism. This new word has rung out like a terrible tocsin; maleficent gases have been seen everywhere, and the *olfactory nerves* have become surprisingly sensitive."[16] He also pokes fun: "Parisian frivolity will much enjoy seeing chemists decant air like thimbleriggers and then bring their olfactory nerves to bear on mephitized lavatory seats."[17]

Evidence of the new sensibility abounds, especially with regard to excremental odors. Scandals arose in Paris over the old way of clearing cesspools, that is, without ventilation and with badly joined tubs and barrels. Quarrels proliferated between cesspool clearers and neighbors.[18] Clearing a cesspool was considered "dreadful torture."[19] When a repair proved necessary "the inhabitants of the house led an uneasy existence" the whole time the work was being done.[20] Passersby complained. Henceforth cesspool clearing was a public issue. Lavoisier, Fougeroux, and Milly, delegated by the Académie Royale des Sciences to test the new processes, assembled various people in order to consult them on variations in odor. The refuse dumps of Montfaucon were beginning to arouse indignation.[21] The inhabitants of the faubourg St.-Martin and of the rue de Bondy protested in 1781.[22]

Mud also assailed the new sensibility. "One might think," wrote J.-H. Ronesse in 1782, "hearing the complaints that are mounting daily, that the roads were always clean in the past. However, the truth is that people did not even think of complaining in earlier times."[23] The new fashion, which prescribed walking, stimulated indignation. On Tronchin's advice, even aristocratic ladies left their carriages, mephitized by crowded bodies, to fill their lungs with air, which they henceforth required to be pure.[24]

Damours noted that people living in the neighborhoods evinced a new intolerance of slaughterhouses and tallow-melting houses.[25] But only after 1750 did the hygienic status of Les Halles come to the forefront of public consciousness.[26]

The foul odor gave rise to polemics; the fetidity of cesspools, wells, fouled walls, drains provoked anger; "for some years people have been more concerned than in former times with averting the dangers that we have to fear from certain vapors . . . this has given birth to an infinite number of disputes, grudges, and lawsuits."[27] The new sensibility spread from the top to the bottom of the social pyramid in what is now a well-known process.[28] As discussed earlier, the chemists had suggested a systematically ordered set of images of the healthy and the unhealthy largely determined by the feasibility of olfactory analysis.[29] In the field of hygiene "what has been and has in no way changed, has suddenly become unbearable."[30] Medicine, for its part, offered only the vague etiological figures of a discourse which was hesitant, polymorphic, consumed by doubts, and which thereby perpetuated the confusion between miasma and stench, the evil-smelling and the unhealthy, the mephitic and the suffocating. As a result of the vacillating character of medical discourse, the whole question of the sense of smell became an emotionally charged one; the collective imagination was both obsessed and confused by what seems to us to derive more from phantasm than from scientific theory.

Thus the masses' great fear of hospital and prison, recently illuminated by Michel Foucault, intensified, all the more so because, for the masses, every danger became apparent through the senses.[31] Dominique Laporte offers a different explanation. Influenced by Lacan's thought, he argues that the slow formation of a strong centralized state allegedly instituted a new experience of the sense of smell. Henceforth "apprenticeship in smelling [was] directed entirely toward excrement."[32] The odor of excrement came to be perceived as intolerable, and at the same time the treatment of it became private. Since every odor was referred to the odor of the feces, the edict of Villers-Cotterêts, which ordered every individual to "look after his own shit," made smell tend to disappear.

The primary event in this prehistory of the olfactory revolution was linguistic. In the seventeenth century, classical French was purified, cleansed of the more malodorous features of its vocabulary. The result was an initial decline in references to smell and an "obscene twisting of syntax" when the mention of excrement was unavoidable.[33]

The olfactory revolution was clearly accomplished via the chronicles of the cesspool, the flood of discourse devoted to filth with a view to abolishing it. But this malaise of hypersensitivity could be only temporary, since it implied the creation of that deodorized environment—our own—made possible by the creation of a strong state that initiated a new way of administering excrement. How that development may have governed the process is an exciting question; but how coherently it stands up to investigation is a matter we shall leave to others.

On the other hand, the privatization of human waste was only one aspect of a larger trend, the rise of the concept of the individual.[34] All the evidence suggests that this concept played an important role in the rise of intolerance in exactly the same way as the new "spatiality of the body" emphasized by Bruno Fortier.[35] Menuret cited as traditional behavior the revulsion against "the atmosphere of people" experienced at times of plague.[36] The fact that the odors of the "I" were better defined, more intensely felt, could only stimulate repugnance to other people's odors, the bodily odors of the rich rotting in the churches, the odor of the crowd sweating in confined areas of public space.[37]

"Odors" and the Traditional Therapeutic Defense

In the mid-eighteenth century the profusion of aromatics was still a contributing factor in the strong smell of the environment. The therapeutic function of "odors" strengthened their aesthetic, or at least their hedonistic, value.[38] Moreover, very little distinction was made between the two roles. "Wearing a pleasure perfume" or burning sweet-smelling pellets in the perfume-pan was deemed to stem infection.[39]

Aromatics and perfumes, as well as certain foul odors that also were thought to have therapeutic value, had an important place in pharmacopoeias, as attested by Lémery's *Pharmacopée universelle,* published in 1697 and regarded as authoritative long afterward. A century later, Virey devoted two long scientific papers to osmotherapy.[40] Lorry's classification of odors in about 1783 had a therapeutic purpose.[41] Belief in the virtues of perfumes was rooted in antiquity: eighteenth-century doctors referred to Hippocrates and Galen, of course, but even more to Crito, whose therapeutic measures, according to Aetius, were based entirely on the use of aromatics.

The nose's proximity to the brain explained the speed and forcefulness with which inhaled odors acted. Lémery suggested the prescription of highly odoriferous "apoplectic balms" because "what is pleasing to the nose, being composed of volatile, subtle, and penetrating parts, not only affects the olfactory nerve, but is spread through the whole brain and can deplete its pituita and other overcoarse humors, increasing the movement of animal spirits."[42] A century later Banau and Turben stated that for the same reason it was more dangerous to inhale a mephitic substance through the nose than to breathe it through the mouth. The proximity of the brain increased the risk of death by stroke.[43] This proximity also enabled odor to rejoice or sadden the soul according to circumstances. Its effects on psychic states thus justified the "medicine of vegetable spirits," which was intended to correct possible disorders in the circulation of animal spirits.[44]

For seventeenth-century mechanists and their followers odors also had a mechanical effect on the organism. Inhaled through the olfactory passages or taken in through the vagina, aromatic effluvia provoked or "destroyed the vapors" of the womb. "It is claimed that civet, musk, and ambergris applied to the navel and toward the womb, when it has been shaken at a time of vapors and choking, affect the womb below by their pleasant odor and restore it to its natural condition, in the same way that these odors agitate and raise it when they are received through the nose."[45] Lémery was merely describing one of the practices of ancient medicine.

Aromatics had a double virtue in respect of the risk of contagion and infection: they fought corruption in the atmosphere and increased the organism's resistance. Odor could restore the elasticity of air and deprive disease of its sting.[46] In this respect medical writing remained hazy; usually it sustained the confusion between the loss of a physical quality and the presence of a possible miasmic burden.

Blégny and Lémery, like the doctors of the period, were convinced that aromatics could restore vitiated air.[47] "Perfumes"—used here in the sense of fumigants—were capable of destroying the deadly presence of the plague lodged in porous bodies, fabrics, clothing, and bundles of merchandise. This was the justification for the disinfection techniques used in Mediterranean lazarettos well into the nineteenth century,[48] despite the interminable quarrel they provoked between contagionists and anticontagionists.[49]

In about 1750, doctors looking for antiseptics to offer effective resistance to the degenerative effects on the humors of certain putrid miasmas, were induced to justify scientifically the therapeutic quality of certain aromatics. All this was later discredited by pneumatic chemistry. As discussed earlier, Becher thought that odoriferous substances facilitated the circulation of the balsamic spirit of the blood and thus slowed the advance of putrefaction. The discovery of aeriform exchanges between living organisms and their environment led to the idea that because of their volatility aromatics naturally became distributors of "fixed air."

According to Pringle, camphor, serpentaria, flowers of chamomile, and quinine, all odoriferous substances, were the most effective antiseptics.[50] Lind recommended camphorated vinegar or odoriferous resins.[51] French doctors echoed British experts. "Putrid exhalations," wrote Boissieu, "will be corrected by the boiling of vinegar and the burning of aromatics several times a day."[52] Gardane proffered the same advice. Bordenave refined the analysis and diversified aromatics' field of action.[53] Some odoriferous antiseptics were stimulants or tonics, which increased resistance to putrid infection; others were astringents, which helped reduce the organism's exposure to infection; the balsamics corrected the consistency of humors already affected by putrefaction.[54]

However complex and fragile its theoretical basis, the belief in the qualities of aromatics governed actual behavior. "The aromatized man" corrected his atmosphere by strong smells where necessary, by the heavy exhalations of musk, ambergris, or civet. To use excessive amounts of perfume was to protect oneself and to purify the surrounding air. Thenceforth there was nothing surprising about the long-lived fashion for animal perfumes with an odor of excrement; only Louis XIV's authority seems to have been capable of checking it for a time, at least at Versailles.[55]

Traditionally, in a period of epidemic people tried to protect themselves by covering themselves with aromatics. Papon summarized these old practices in 1800: "A sponge soaked in vinegar or a lemon studded with cloves or an odoriferous ball will be carried in the hand and sniffed from time to time. For people who are not in a position to afford odoriferous balls or perfume-pans, the best authors recommend sachets of rue, melissa, marjoram, mint, sage, rosemary, orange blossom, basil, thyme, serpolet, lavender, bay leaves, orange and lemon bark, and quince rind; they recommend that these always be present in apartments at time of plague."[56] Buc'hoz recommended sniffing red carnations and sprinkling clothing with pulverized angelica.[57]

For a long time the best protection against disease remained possession of a shield against smells, smelling strongly oneself, and also sniffing odors of one's choosing. It was therefore a good idea to have a "smell box" in one's pocket, stated Lémery.[58] Lind recommended wearing camphor in an amulet and perfuming one's clothing with it.[59] Guyton de Morveau copied the health officers around him and provided himself with a flask of vinegar.[60] Baumes noted that many people were accustomed to soak little sponges in camphor and to raise them "to their nose or mouth every second";[61] he recommended that workers employed in clearing marshes be made to sniff them in exactly that fashion. Ramazzini advised gravediggers to carry cotton wool impregnated with vinegar on their person and "to breathe its odor from time to time to restore their sense of smell and their spirits."[62] Fourcroy issued instructions to stoneworkers: "They should go down to their quarries only if provided with a sachet hanging from the neck, containing two cloves of garlic pounded with a little camphor. They will rub their faces with camphorated brandy or aromatic wine."[63] In the mid-nineteenth century, well after the advance in medicinal chemistry, people continued to arm themselves with sachets dispensing healthy protective odors. Parent-Duchâtelet made them compulsory for workers employed in clearing the Amelot sewers in 1826.[64]

Aspersion or fumigation was thought to disinfect the surrounding air. The masses had more confidence in hot vinegar than in anything else. Surprisingly to us, the odor of vinegar was considered balsamic.[65] Sulfur, gunpowder, and sealing wax were also burned; even more, aromatic woods, rosemary, and juniper berries; bottles of scented water were used for fumigation.

There were many methods of fumigation. The most common was to pour vinegar onto a shovel reddened by flame.[66] More refined was the use of pellets or troches placed on hot ashes. The use of perfume-pans, especially silver ones, signified membership in the

elite. Master perfumers still prepared Bruges strips, specially intended for fumigation. The supreme luxury was the *parfumoir,* "a
small wooden chest equipped with a grille. At the bottom . . . is a
small opening for the insertion of a chafing dish full of fire where
pellets are put to burn."[67] In lazarettos, mail from the contaminating
Orient was "disinfected" in this way.

Before the triumph of scientific methods of fumigation by means
of chemicals, disinfection tended to transform the surrounding air
into a multitude of smells.[68] It increased the strong smell of an already
badly ventilated dwelling. The odors of juniper and rosemary were
reserved for sickrooms. Fumigating the ground floor was enough to
disinfect the whole house: the rising smoke invaded the upper stories.
Chests and cupboards were emptied, and clothing was hung out so
that it could best be impregnated by the redeeming odors.[69] During
the great plague in Marseilles in 1720, disinfection teams carried out
three successive fumigations, "the first with aromatic herbs; the second with gunpowder; the last with arsenic and several other drugs
that have been used in lazarettos since time immemorial."[70] Père
Léon's perfume, the "vinegar of the four robbers," worked wonders
at that time.[71]

Once again ships and hospitals provided the models. The elder
Lind, a Portsmouth doctor, was the first to codify the disinfection of
places that had been made putrid by overcrowding.[72] Tenon reported
that he ordered not only that clothing be fumigated but also that
large quantities of powder be burned between decks and in the holds
of contaminated vessels, and "a great smoke" spread there. On the
Continent, Morogues refined Lind's advice: "It should be possible
to send aromatic vapors between decks by having a red-hot iron
spoon carried along there, containing resin or tar or juniper seed or
powder moistened with vinegar or other inexpensive aromatics, and
thrown in, a small quantity at a time."[73] His prescriptions were followed.

All observers attest the use of aromatic fumigation in hospitals;
some emphasized this practice, if only to deplore it, until well into
the nineteenth century. Here, as in private homes, juniper and rosemary predominated.[74] Fumigation with incense and storax in churches
was seen as an act of worship, but it also tended to cause the stench
from the corpses underground to be forgotten; experts saw it as a
powerful means of disinfection and an effective protection against
the putridity of the assembled worshipers.

In this period of epizoism, aromatic vapor became ubiquitous, penetrating even into stables; in criticizing the practice, Vicq d'Azyr bore witness to its universality.[75] Following the example of Hippocrates, who burned stakes to combat the plague of Athens, several doctors nursed the astonishing plan of perfuming the city. After all, rue had successfully protected a whole district of London during the Great Plague of 1666.[76] A century later, the thick smoke from 120 juniper-wood stakes all lit at once in the streets of Bois-le-Roi was enough to vanquish the epidemic.[77]

Apart from coal smoke, sometimes feared because of its subterranean nature (especially in England), smoke caused no feelings of disgust until later; for the moment, what was intolerable was the odor of putrefaction or fermentation, not of combustion. Indeed, according to some, the fires of industry in the heart of the city could counteract the emanations of the "great unwashed," the vapors from refuse, and the inherited infection of the soil from ages past.[78] Thus the discourses on urban salubriousness were not unanimous; the ecological dream has some surprising byroads; and anachronistic readings from later attitudes are to be avoided.

Using odoriferous substances for fumigation was also part of the therapeutic repertoire, although the fashion for these methods seemed to be on the decline, except in cases of hysteria. The volatile nature of the "perfumes," their powers of penetration, the mysterious rapport that was established between nose and womb, led doctors to use the "odors" as antispasmodics to calm their female patients' attacks. Fumigation with paper, old shoes, and other stinking substances soothed rising vapors and cured amenorrhea. The smoke from headache powders strengthened the brain. Fumigation with astringent mixtures checked the common cold. Apothecaries prepared scented sachets to cheer up melancholics; hypochondriacs' clothing was perfumed with aromatic powders. Fumigation with cinnabar cured the pox.[79]

The fashion for strong smells and the practice of aromatic fumigation did not suddenly disappear; moreover, the rhythm of their retreat varied from circle to circle. Josephine and the "Merveilleuses" of the Directory revived the fashion for musk; the use of balsamic vapors sometimes reached feverish proportions, fanned by the great nineteenth-century epidemics. Yet strong odors had been denounced and disinfection by aromatics criticized for nearly a century. It is important to analyze this theoretical devaluation.

The Discrediting of Musk

In the late seventeenth century, from Becher's viewpoint as from the ancients', excrement was endowed with vital fire and so had a therapeutic property; there was therefore nothing aberrant in using it in aromatic preparations, notably in the composition of Eau des Mille Fleurs, all the more so when the fecal matter came from strong and healthy individuals. Beginning in the mid-eighteenth century, however, the harmful effects attributed to putrefaction revolutionized attitudes to the products of defecation and, in a more general way, to all the animal substances hitherto used in perfumery.

Pringle's and MacBride's experiments may have temporarily increased the value put on aromatics, but they also led doctors to regard musk, ambergris, and civet as putrid and eminently septic substances; their excremental component was constantly reiterated and exaggerated in order to emphasize their noxiousness. Scientists criticized the dangerous affinities between the stifling smells of these particular aromatics and the smell of excrement.[80] Boyle had already argued that musk which had lost its odor "regained it and was restored by being hung over a humid floor for a period, and particularly near a privy; which means that the nature of musk is recremental"; he had also noted that stables and sheepfolds smelled of musk.[81] Virey declared that if human excreta were digested and fermented in a double boiler, they acquired a musky odor;[82] Friedrich Hoffmann attributed this odor specifically to bile. Ruelle claimed that the same was true of rats' excreta, which he accused perfumers of using to doctor their products. Hartley stated that dung smelled of musk when one stood a few steps away from it. The preparation of Eau des Mille Fleurs became the target of chemists and sanitary reformers. The odor of the musk deer's abdomen was said to have killed the foolhardy hunter who failed to hold his nose before approaching his prey.

Another argument, already advanced by Boerhaave against strong odors, was that they exhausted the psyche, created or revived anxiety, sometimes gave rise to stupor. The first sensation might be pleasurable, Buffon confirmed, but pain came at the end;[83] in the realm of olfaction, a threshold of pain separated sweet scents from overly powerful perfumes. Headaches were the least of the ills; aromatic odors themselves could cause "intoxication" of the sense of smell, however tonic and stimulating they were at the outset.[84]

Madame de Sévigné was for a time fascinated by the beneficial effects of Eau de la Reine de Hongrie and felt obliged to warn Madame de Grignan against abusive inhaling of what had for her become a positive drug.[85] According to Lorry, musk deranged female nerves and ravaged male stomachs.[86] Fourcroy, quoting Bacon and Ramazzini, noted that apothecaries and their assistants were prey to terrible accidents.[87] Midwives who were forced to equip themselves with heavy perfumes as protection against putrid emanations from women in labor were accused of making their clients hysterical.[88] Nor were animals immune to the disastrous influence of violent perfumes; on one occasion mules transporting saffron fell into a faint.[89]

More serious effects were also possible. Numerous anecdotes attested that heavy scents concealed devastating poisons. Henry VI died from sniffing perfumed gloves, Pope Clement VII from going too close to an odoriferous torch. It was recalled, though not seriously credited, that an Indian queen had thrown into Alexander's arms a magnificent young girl whose breath had been literally poisoned by her habit of sniffing pernicious drugs. There were also innumerable accounts concerning the ravages of hellebore, henbane, cantharis, magnolia, and manchineel.[90] It needed all Orfila's authority half a century later to relegate odors to the status of a partial poison. All these references are enough to explain why a guard was hired for a woman in childbirth to protect her from importunate visitors steeped in deadly scents.[91]

Advances in bodily hygiene among the elite stimulated distrust of offensive odor. The use of powerful perfume cast doubt upon a person's cleanliness. Musk aroused suspicion. The same was true concerning public space. Howard blamed aromatic fumigations for concealing negligence inside hospitals.[92] On the other hand, the advance of the *toilette de propreté* encouraged the fashion for subtle and delicate odors. It was dangerous for people who washed and bathed naked to use violent perfumes. Careful choice of toilette odors that "rapidly penetrate the whole animal economy through the absorptive system" was therefore important.[93]

After 1750, the use of heady odors also suffered from the fashion for the natural, which argued that the odor of the flesh, simply enhanced by fragrant floral effluvia, should be allowed to filter through the filmier toilettes of the day. Finally, exciting perfumes—together with balsamic odors—were implicated in the criticism of luxury and artifice.[94] According to Pluquet, Abbé Jacquin thought that vinegar, sulfur, and gunpowder were the only salubrious "perfumes." He heaped anathema on aromatics and abused courtiers who "wore odors." His criticism was moral rather than scientific: "the odors," he argued, "belong less to cleanliness than to a certain depraved taste or air of fashion."[95] According to Caraccioli, perfumes contributed to the confusion of sensory impressions, a damning effect: "as if the nose ought not to be satisfied with smelling, the eyes with looking, and the tongue with tasting." Wine was aromatized; tobacco emitted effluvia of jasmine; sugar was flavored with ambergris; everything eaten was perfumed. "The five senses, confused in this way, have become the soul of sensualists and they would not wish to acknowledge any other."[96] The criticism of odors was part of the much wider criticism of artifice, affectation, effeminate fashion—in short, all the tendencies suspected of leading to "degeneration."[97]

A fundamental aspect of this revolution in olfactory tolerance was the linkage forged between the criticism of "odors" and the rise, and then the spread, of the bourgeois mentality. According to its etymology, perfume was dissipated in smoke. What disappeared or became volatile symbolized waste. The ephemeral could not be accumulated. The loss was irremediable. One could dream about recovering and reutilizing waste or about recycling excrement; evaporation was beyond hope. For the bourgeois there was something intolerable in this disappearance of the treasured products of his labor. Perfume, linked with softness, disorder, and a taste for pleasure, was the antithesis of work. Aside from its possible therapeutic function, it had no "secondary utility";[98] it was immoral. All in all, it was desirable that it lose its animal references and that its exciting allusions to the reproductive instinct disappear together with musk.

The Denunciation of Aromatics

Shortly after doctors and moralists had stressed the dangers of animal perfumes, pneumatic chemistry began to challenge the therapeutic value of "odors" and "aromatics." Paradoxically, it seems to have been MacBride who challenged their protective property scientifically. In France as early as 1767, Genneté was arguing unanswerably that perfumes did not augment phlogiston; on the contrary, they destroyed it.[99] In 1775 Vicq d'Azyr denounced aromatic fumigations as totally ineffective.[100] Potpourris and voluptuous perfumes were unsuitable for reviving the elasticity of air, declared Abbé Jacquin.[101] But it was Guyton de Morveau who formulated a theory with which to discredit aromatics.

Fumigation with aromatics was ineffective because it did not produce any transmutation. Real disinfectant had to destroy certain preexistent substances and bring into being new bodies in the environment that were amenable to chemical analysis.[102]

This conviction was soon shared by most experts.[103] "Let us make haste to proscribe perfumes," clamored A.-A. Parmentier and J. A. Chaptal: "the fumigations with incense, etc., that are in common use do nothing but mask unpleasant odor."[104] Advances in medicinal chemistry, illustrated by Malouin's treatise in particular, sanctioned and stressed the therapeutic devaluation of aromatics. During the same period Ingenhousz demonstrated that respiratory exchanges in vegetation were unaffected by the fragrance or fetidity of the odor exhaled.

Nevertheless, the analyses of Priestley and Ingenhousz could not explain the undeniable influence of the odor of certain plants, any more than analysis of the air could detect miasmas. In these circumstances some scientists, failing to purify vitiated air, thought that perfumes were perhaps not in fact totally neutral. From this context came an increasing belief in the beneficial effect of spring flowers, deemed to be bursting with life. Their perfume was accepted as the antithesis of the putrid and excremental odors that had to be shunned. Fourcroy denounced musk, condemned artificially perfumed air in apartments, but glorified inhalation of the natural and balsamic air of meadows.[105] Rousseau's influence on this notion is obvious; yet we cannot be sure that his depiction of Julie's garden was not merely reflecting the already established direction of contemporary medicine.

Hallé's official code, published in 1818, registered the break with traditional attitudes but still conveyed an ambiguity of beliefs and behavior.[106] The skepticism of the scientists toward traditional beliefs has become glaringly visible. The authors of the code confirmed the loss of confidence in the effectiveness of aromatic fumigation, denied the therapeutic property of odors, and sounded the victory of chemical medicaments. However, they did not feel authorized to gainsay practices that were still deeply rooted. They tolerated the most fragrant aromatics in the form of compound aromatized spirits, often distributed under the name of elixirs; they encouraged the use of perfumes in pharmaceutical preparations. Overall, they tended to relegate odoriferous substances to the role of secondary aids, and thus caused the confusion between pharmacy and perfumery to be perpetuated.

5 *The New Calculus of Olfactory Pleasure*

Pleasure and Rose Water

AMONG THE ELITE, changes in tastes and in fashion sanctioned the experts' discrediting of heavy scents. The smells of private space became less strong and were enriched and varied by more delicate and subtle fragrances. The new behavior patterns reflected the fascination with airy space. The balsamic effluvia of springtime meadows became an obsession. The interiors of Tiepolo's paintings corresponded to the inarticulate expression of a new sensitivity to smell. Lowering the thresholds of perception not only created intolerance of excremental odors; it also emphasized the social function of personal toilette as an aspect of good manners, which were being codified in an increasingly strict and precise manner.[1] Intrusive perfumes as well as indiscreet body odors had to be avoided lest they cause discomfort.

Ernst Platner, the authority most commonly referred to in the late eighteenth century, listed the theoretical dangers of bodily uncleanliness. Dirt obstructed the pores; it held back the excremental humors, favored the fermentation and putrefaction of substances; worse, it facilitated the "pumping back of the rubbish" that loaded the skin.[2] This foul-smelling pellicule, too often regarded as a protective coating against miasmas, impeded the aeriform exchanges necessary for organic equilibrium. It was therefore important to increase the frequency of ablutions. Platner, like Jacquin, recommended that faces, hands, and feet be washed often, and even the whole body "from time to time."[3]

In this way a very cautious, somewhat uncertain bodily hygiene was encouraged, its spread limited by a multitude of restraints. Vitalists and iatromechanists appealed for prudence. The loss of vitality that Bordeu stressed was not the only danger from water. Rash use of baths relaxed the fibers, weakened the organism, led to indolence. Like Boyle and Lancisi not long before, Hallé emphasized the septic effect of soap, particularly in time of plague.[4] But moralists were afraid of the pleasure, the sensual glances, and the autoerotic temptation of baths. Privacy was provided as a proof against seduction in the dressing rooms of the period;[5] nakedness was risky.

In any case, such practices were bound to be confined to a small elite. Inadequate control of running water prevented private bodily hygiene from becoming widespread.[6] For most people, the use of water remained collective.[7] Bathing, which did in fact become more widespread at the end of the century, at least in Paris, was more a therapeutic practice than anything else.[8] Furthermore, Moheau noted, ablution was useful for the common laborer only when he was not working; for the rest of the time the movement of his sweat was enough to clear his pores.[9]

Nevertheless a popular pedagogy of private hygiene was sketched out; norms were formulated within limited circles—the same circles that gave rise to the keenest anxiety; schools and, even more, prisoners' cells, hospitals, barracks, and Cook's ship became the laboratories for little-known experiments.[10]

Among the elite, the new use of perfumes coincided with the new ritual of the toilette: the individual must not betray poor hygiene by wearing a scented mask. Quite the contrary, the individual atmosphere revealing the uniqueness of the "I" must be allowed to break through. Only some vegetable odors, chosen with discernment to express a certain olfactory harmony, could enhance allurement of the individual person. The woman developed a wish to breathe and control her fragrances at the same time that she began using the looking glass. The psychological and social function of delicate scents justified the new fashions. "It is necessary to do something to make us pleasing to ourselves," wrote the perfumer Déjean concerning the use of vegetable perfumes. "That makes us lively in gatherings and in that way we please others; that is what makes society. If by some misfortune we are displeasing to ourselves, whom will we please?"[11] This comment confirms a development of the greatest importance, which Roger Chartier has already stressed with regard to school manuals: the movement away from a code of good manners primarily intended to avoid causing embarrassment to other people, toward a body of hygienic precepts that also aimed at narcissistic satisfaction.[12] Woman wanted to be breathed; she thereby affirmed her wish for self-expression. By this discreet allusion to the body's élan, by this search for an image, she created an aura of dream and desire. One might say that the shift from a "mosaic" to a "syntactic" model, which was to underlie the emergence of high perfumery at the end of the nineteenth century, had been prepared for much earlier.

The new fashion, all subtlety and delicacy, translated the major historical shift discerned by Robert Mauzi: the movement from provoked sensation to received sensation, from artifice to nature.[13] What unleashed sensual agitation was the vagueness of the soliciting message. "Perfumes are used," wrote Déjean, "to satisfy the sensuality of the sense of smell . . . not with those strong and violent odors, but with those sweet odors which can be neither distinguished nor defined."[14]

The application of these principles led to the rejection of animal perfumes. Ambergris, civet, and musk went out of fashion, "since our nerves have become more delicate," according to the 1765 *Encyclopédie*.[15] Musk-scented gloves were no longer tolerated; their odor was too violent. There is a vast amount of evidence on this subject. Musk had become old-fashioned, declared Nicolas Le Cat.[16] Déjean alluded to the disrepute into which this perfume had fallen as if it were self-evident, and was content to plead the cause of ambergris.[17] Nevertheless there is some evidence of more conservative attitudes. Whereas animal perfumes were the subject of abuse, a belated craze developed for "extracts of royal ambergris." The resistance, contested but indisputable,[18] is explained by the interplay of taboo and desire, a subject that demands special attention.

Havelock Ellis rightly analyzed this discrediting of musk as a major fact in sexological history. In his opinion, up to the end of the eighteenth century women did not use perfume to mask their odor but to emphasize it.[19] Musk had the same function as corsets that accentuated the contours of the body. According to Hagen, master of sexual osphresiology, women until then had sought out the strongest, most animal perfumes for that purpose.[20]

From this viewpoint, the decline in these perfumes at the end of the eighteenth century only registered the decline in the "primitive attractiveness" of sexual odors.[21] Havelock Ellis took up Bordeu's timid analyses. Western man and woman would henceforth endeavor with increasing skill to disguise body odors that had become burdensome; it was a way of denying the sexual role of the sense of smell, or at least of shifting the field of olfactory stimulation and allusion; thereafter it fell to the delicate exhalations of perspiration, and not to the powerful odors of secretions, to presage the intimate liaison. Never before had there been so great a revolution in the history of sexual invitation; except, Freud wrote twenty-two years later, when man stood upright and thus for the first time reduced the role played by olfaction in unleashing desire.[22]

There is nothing to oppose the idea that sensualism contributed to the victory of that prohibition on animal perfumes. The excremental odor of the emunctories situated near the genitals—as is the case with the musk deer's abdomen—would explain the sense of shame and, in the final analysis, the feelings of modesty that the genitals aroused. Hartley was convinced of this connection: "The mental displeasure that accompanies shame, the idea of indecency, etc., come to a considerable degree from the disagreeable odors of fecal substances of animal bodies."[23] Thus the English philosopher justified an idea dear to the fathers of the Church; his theory led implicitly to a condemnation of the use of musk, ambergris, and civet.

We have already considered, in all its strange complexity, the theoretical justification for the decline of animal perfumes. It was accompanied by an enormous vogue for "odoriferous spirits," "essential oils," and "scented waters" extracted from spring flowers. What was new was their great variety.[24] At the court of Louis XV, etiquette prescribed the use of a different perfume every day. The great success of rose water[25] was shared by violet, thyme, and, above all, lavender and rosemary water. "Lavender water," noted Malouin, "is the most generally pleasing of all the odors."[26] The 1760s saw the launching of "eaux de la maréchale" and "eaux de la duchesse," whose fashionable image reinforced sensitivity.[27] A few years later vegetable odors "from the islands" contributed an exotic note to the range of floral essences.[28] Men as well as women obeyed the new injunctions; Casanova made fun of the young Baron Barois, whose chamber smelled sweetly of the odor of the pomade and scented waters he applied.[29]

Delicate "odors" became part of the ritual of bodily hygiene. Of course, several doctors, Platner foremost among them, advised the use of pure water and avoidance of perfumed mixtures.[30] They were barely heeded. "L'eau d'ange" was much in vogue in the seventeenth century, but according to Déjean its use had lapsed by 1764. It was followed by waters with fruit scents, soap and paste with floral perfumes, odoriferous balls to rub on the body in the bath. Master perfumers prepared scented pellets and powders to scent the hands—a subject that aroused lively interest.[31] The custom developed of washing out the mouth with rose water, perfuming the breath with iris paste.

Courtly literature was quick to record the discrediting of musk. Notions of hygiene and ablutions are central to Restif de La Bretonne's eroticism. Rose water had a surprising monopoly; it was ceaselessly refreshing Conquette-Ingénue's feet and private parts.[32] The bidet became the accessory of pleasure. Casanova's story has the same monotony as far as the use of scents is concerned: washing the woman's body in rose water assumed the form of a ritual.[33] Perfume played hardly any role except in setting the scene for pleasure;[34] it had moved away from the desired body. It would even disappear from the erotic scenario of the marquis de Sade's novels.[35]

The persistent emphasis on the skin's absorptive power justified the caution regarding the use of strong scents. This caution, however, was counterbalanced by the use of perfumed powder, which, more than other cosmetics, revealed the personality of the person using it. It "varies," Déjean noted, "according to each person's taste and is composed of distinctive perfumes."[36] Powder "à la maréchale" retained its prestige for nearly a century; it consisted of a skillful mixture of iris, cloves, lavender, rose, orange, and marjoram, perfected by the maréchale d'Aumont. Apart from this mixture, the ones most commonly used were iris powder, cypress powder, and, above all, carnation powder. Carnation powder came to the fore in the reign of Louis XV;[37] its success signified the triumph of vegetable scents.

A predilection for flowers quite logically accompanied this craze. Fashionable Parisian ladies grew cloves and basil in pots.[38] Large vases of flowers decorated ladies' toilette tables. Ladies of fashion draped themselves with honeysuckle and wildflowers: buttercups, hyacinths, jonquils, lilies of the valley, convolvulus, and ranunculus. A veritable cult surrounded violets. Marie Antoinette consolidated and extended a set of behavior patterns that had appeared well before she imposed her own image on the court.[39]

Strong odor, now old-fashioned, became the prerogative of aged coquettes and peasants. Animal scent belonged to the masses. "The gentleman of fashion no longer exhales ambergris," noted Louis-Sébastien Mercier.[40] Casanova nearly fainted when an old nymphomaniac duchess made her appearance, smelling of musk from twenty steps away.[41] He himself used only myrrh and storax to concoct the magician's sulfurous apparatus.[42] The captivating Célestine nevertheless rejected Casanova by mocking scented waters and revealing her peasant origins through the avowal that she used goat's fat.[43]

"Apart from philosophers . . . everyone smelled nice," Alexandre Dumas wrote in 1868 of the elite at the end of the ancien régime.[44] Edmond de Goncourt and Huysmans helped to establish the myth that the eighteenth century was flower and perfume mad. There is an element of truth behind this myth, which the contributors to the *Encyclopédie* recognized. The practice of using perfume to embellish the environment and the objects in it acted as a counterweight to the disappearance of musk and civet. Perfumers offered "concoctions to wear on one's person," "perfumes for pleasure," with no therapeutic purpose.[45] "Odors are carried only in bottles," Déjean specified, "for fear of annoying those who do not like them."[46] Cotton wool soaked in perfumes was concealed in miniature perfume-pans or in tiny tassels sewn onto an item of dress. Gentlemen of fashion vied in the art of analyzing the concoctions. Possession of a royal perfume was a sign of membership in the aristocracy of refinement. Casanova took good care of his bottle from the suite of Louis XV.[47] Sade dispatched persistent pleas to his correspondents to send him rich "odors" when he was imprisoned in the Bastille.[48]

The perfumed handkerchief, a favorite component of feminine wiles, remained fashionable into the nineteenth century.[49] Both perfumers and coquettes made up "pads to wear on the person" and filled them with lightly musked violet powder.[50] "English" sachets in Florentine silk or taffetas—because linen corrupted the odors—were fabricated at home.[51] Containing a small layer of perfumed cotton wool or a pinch of fragrant powders, they were attached by ribbons to negligees or were placed in cupboards, chests of drawers, and the drawers of bedside tables.

It was also the custom to perfume toilet accessories. Delicately scented Provence gloves replaced musked gloves.[52] Perfumed fans spread and moderated scents from breasts and bouquets. The harmony established by the odor of the perfumed glove was a sign of the subtlety of the composition of the scents. "English" or "Montpellier" styles of dress cut in odoriferous fabrics were less common.[53] The wearing of negligees that had been kept in delicately perfumed containers confirmed the change in fashions of sexual solicitation.[54]

Everything worn, even medals and rosaries,[55] could caress the sense of smell. Forbidden to smoke in the presence of ladies, men used snuff smelling of jasmine, tuberose, or orange blossom.[56] Cooks busily perfumed their dishes.

The atmosphere of private space was tinged with delicate odors: perfumed boxes, baskets of scent, and, above all, skillfully prepared potpourris, some of which kept their odor for ten or twelve years, perfumed the apartments of the rich.[57] Their concoction, like the concoction of pomades, powders, and scented waters, fostered a domestic art of preserving odors that competed with the master perfumers'.

This moderate advance in bodily hygiene made the dressing room one of the sanctuaries of seduction. There, as in the adjoining boudoir, the smell of the environment, combined with the effect of hangings and mirrors, lent itself to the transactions of intimate exchange. Parny, following Rousseau, dwelled reverently on this favored place for sexual invitation, epitomized by Pompadour.[58] On the other hand, the extreme example supplied by the duc de Richelieu, who had odors skillfully wafted about in his apartments to set the stage for seduction, does not seem to have caught on.

The Perfume of Narcissus

Providing an accessible locale that would allow both the chosen sensation to be welcomed and the pleasures and feelings it aroused to be experienced was the first injunction of the sensualist ethic. Rousseau made this art of sensations, based on the choice and arrangement of objects, the first technique of happiness. It was a difficult calculation, involving constant care to avoid obtrusive sensations that created distractions or even aroused repulsion. Enjoyment of the true pleasures of the sense of smell therefore presupposed a flight far from mud and dung, far from the putrefaction of living bodies, far from the confined places of the town as well as from the cramped lands of the valley. Even in the countryside, flight was necessary; the village had become a cesspool, declared Girardin.[59] "I see a hundred crowded thatches," deplored Obermann, "an abominable cluster where roads, stables, and kitchen gardens, damp walls, floors and roofs, even clothes and furniture, all seem to be part of the same slime. Where all the women bawl, all the children weep, and all the men sweat."[60]

Thus was the repugnance to "social emanations," still vague but intensely felt, displayed.[61] Ramond de Carbonnières, who did so much to spread the fashion for mountains, thought that the "commerce in emanations" took place only on the horizontal plane;[62] it was limited to the social activities of the people of the plain or valley. The elite should be able to escape it by taking to the heights, leaving the stenches of congestion to the confined masses.

The rich man had to enjoy pure air. The wide casements of his dwelling, the unrestricted space that surrounded him were not enough. Tronchin advised him to take walks for fresh air and warned him against the stagnation of repose. Diderot and Sophie Volland left Paris every summer, one for Chevrette or the Grand Val, the other for Isle.[63] Saint-Preux was surprised that vaporish people were not sent to the mountains. In 1778 Thouvenel endeavored to promote aerotherapy, still in its early stages and made fashionable by the philosophers. Jurine advocated "aerial baths."[64] The "air cure," still a vague idea, became a medical prescription[65] before public health experts in the next century improved its practice and expertly adapted it to age, sex, and temperament.[66]

Remedial properties were attributed to gardens and mountains, the antitheses of putrid places. Géraud pleaded for more public parks so that the city dweller could go and release his miasmas under their protective shade.[67] But mountains remained the supreme reference point. Remaining at an altitude might, of course, prove hazardous, Saussure warned his readers.[68] The air breathed in the "confines of the ether is dry and devoid of emanations from the inhabited earth."[69] It was distressing to the reckless tourist. The "degeneration" of the Swiss mountain dwellers, the ugliness of their wives, the cretinism of the inhabitants of the Maurienne all were reasons for prudence.[70] But these lonely places at least permitted the access that was vital to the pleasure of sensation. It was the silence of the mountain pastures that provided the context allowing Obermann to enjoy the sound of the fountain.

The garden's seclusion also permitted the creation of the "romantic situation," which Girardin defined as follows: "without being
wild or unkempt [it] . . . must be tranquil and solitary so that nothing
distracts the soul and it can give itself up entirely to the sweetness
of a deep emotion."[71] Here, the sense of smell was invested with
powerful affects—despite its devaluation by those who regarded it
as the sense of animality.

"The pleasures of the senses must have a basis, or at least a
pretext, in nature so that they do not injure reason," decreed Watelet.[72] This requirement for "landscapes or selected nature"[73] led to
the obsolescence of the skillful combination of perfumes in flower
beds in favor of a very limited range of scents. The odor of newmown
hay was henceforth the supreme balsamic reference point. Louis-
Sébastien Mercier, Ramond, and Senancour, here following Loaisel
de Tréogate, praised its refreshing subtlety.[74] "Around four o'clock
I was awakened by daybreak and the odor of hay that had been cut
in all its freshness by moonlight," Obermann disclosed.[75] Thus the
perfume known as "Newmown Hay" made its name. Henceforth
most references to smells in the literature devoted to the pleasures
of nature were divided among jonquils, violets, and jasmine. Roses,
so highly valued by perfumers, seemed old-fashioned. Strawberries
tended to epitomize the pleasant odor of fruit.

Making the garden the source of olfactory pleasures initially seems
paradoxical. A garden was, of course, primarily, a picture. Its composition was based on "the mechanism of sight."[76] Architects were
clearly guided by the desire to indulge sight and hearing above the
other senses. The English garden provided the opportunity to con-
secrate and enact a hierarchy of the senses in a manner that resembled
a religious litany. Girardin extolled the superiority of sight with its
more immediate, keener, subtler impressions.[77] Hirschfeld summarized the position decisively in 1779: Of all the senses "the sense of
smell, which receives the sweet exhalations of plants and vegetables,
seems to be the last, unless it is joined with the coarser sense of
touch, to experience the refreshing action of the air"; consequently
the artist, "though without neglecting the sense of smell entirely,"
had to "work for the eye and the ear, particularly for the eye. Therefore, the gardener will mainly endeavor to expose the visible beauties
of rural nature."[78] Although these developments diminished the role
of flowers, the latter continued to be used primarily to gratify the
eye; their first function was to carpet the hillside, to dot the meadow,

not to delight the sense of smell. The increased number of theatrical scenes set in "picturesque gardens" confirmed the primacy of the visual. Only the ear, reassuring witness of the purifying movement of wind and, even more, of water, could compete temporarily with the eye in the scale of sensual enjoyment. Neither Thomas Whately, whose influence was considerable, nor Jean-Marie Morel alluded to olfactory pleasure.[79]

Nevertheless, the sense of smell did form part of the sensual palette available to the artist who wanted to vary the production of sensations and feelings. Perfume could help to perfect a strategy of emotional satisfaction. Thenceforth it would be irrelevant to try to analyze too exclusively what originated from each sense; to do so would amount to rejecting the quest for "corresponding percep-tions," which, according to Hirschfeld, were essential if gardens were to be places that sated the senses. "A grove embellished by fresh foliage and smiling prospects is even more delightful when at the same time we hear the song of the nightingale, the murmur of a waterfall, and when we breathe the sweet odor of violets."[80]

Analysis of the discourse on landscapes, parks, and gardens makes it appear that references to the sense of smell involve only a limited number of places, attitudes, and feelings. The sense of smell was solicited primarily to accommodate the wish for repose. The ap-proaches to the house and to the bedroom, the arbors sheltering the bower or resting place, the bed of moss that invited a halt in the depths of the valley, and, in a more general way, "serene regions" called for plants with odoriferous flowers or foliage in the vicinity.[81] No one has defined this subtle imperative—already expressed by Horace Walpole[82]—better than Hirschfeld. The model here was not so much Julie's garden as the bower that sheltered the love of the first couple in Eden's "wilderness of sweets" in *Paradise Lost*.[83]

The sense of smell was solicited wherever running water be-stowed its freshness and encouraged sensory associations. Girardin recommended decorating banks of streams with odoriferous plants.[84] In the heart of the oak wood, near the source, "simple aromatics, wholesome herbs, and the resin from odoriferous pines perfume the air with a balsamic odor that expands the lungs" and invited reverie.[85] Hirschfeld advised strewing flowers near bridges where strollers liked to sit down.[86]

It was also possible to deploy heady perfumes when work and the need for fertilization justified them. Locating a flowery close next to the apiary was a practical measure: "Thyme, lavender, marjoram, willow, limes, poplar grow in profusion there and scent the air one breathes over a wide area. Here the luxury of flowers and perfumes is allowed."[87]

The optimistic identification of the natural with the vital and the wholesome was a way of acknowledging both the attraction exercised by the perfume of growing plants and the odoriferous sensuality of the open air. The penetrating odor of some country flowers could intoxicate; it was an inducement to sexual pleasures, as was suggested by the resemblance between the facial expressions of a woman when smelling a flower and when making love. According to some experts, this ambiguous marriage between the woman and the flower she smelled could end in orgasm.[88] The perfumed arbor, the bower it sheltered, places of solitude, repose, and dreams were easily trans- formed into privileged theaters of vertiginous abandon. The perfume of nature merged with the incense of voluptuousness. The young countess's seduction by the guilty Dolbreuse, like the excitement of the open-air wedding night, owed a great deal to the alliance of natural perfumes.[89] Orange blossom, jasmine, and honeysuckle per- fumed the loves of Sydney and Félicia.[90] The subtle hedonism of plant life surpassed even the skillfully perfumed backcloths con- structed for pleasure by the libertine.

The bulky literature devoted to the "English" garden should not eclipse the continued existence of the flower garden, a perfumed enclosure round the house. Both Girardin and Hirschfeld persist- ently emphasized the bourgeois fashion for this "pleasure garden." Women and, even more, young girls found in it something to stir their sensibilities. They went there to compose their vapors by breathing "sweet, delicate, pleasing, and refreshing perfumes, suited to revive their spirits."[91]

The essential function of olfaction in all these favored places was clearly to encourage narcissism. Far from the *theatrum mundi,* in- formed by the staleness of worldly intercourse, tempted by hermi- tages, grottoes in picturesque gardens, or mountain rocks, the reader of Rousseau's *Rêveries,* Werther's confidences, or Young's *Nights* dreamed of intensely experiencing the existence of the "I." Because it contributed to awareness of the flight of the self, the sense of smell was henceforth the favored sense for the perception of time. The landscape artist had to observe the clock of the smells of nature;[92] there were morning, midday, or evening gardens, and he had to choose among them. If he intended to give particular importance to perfumes, he chose evening scents, because it was undisputed that the exhalation of plants emphasized with particular intensity the flight of day. According to Ramond, this was what made the odor of the Pyrenean fringed red carnation so moving.[93] Olfaction thus played a large role in the endlessly savored theme of the seasons.

What was really new was the power of odors to stir the affective memory; the search for the "memorative sign" (as Rousseau put it),[94] the violent confrontation of the past and present engendered by recognition of an odor, could produce an encounter that, far from abolishing temporality, made the "I" feel its own history and disclose it to itself. Just as the growing fashion for subtle perfume gave the remembered image of the Other a poetic resonance, so literary descriptions of smells concentrated on their power to engender reminiscence itself. Two examples will suffice among the thousands available.

"There is something indefinable in perfumes that powerfully awakens memory of the past. Nothing so much recalls beloved places, regretted situations, the passing of those minutes that left as deep imprints on the heart as they left few in the memory. The odor of a violet restores to the soul the pleasures of many springs. I do not know what sweeter moments in my life the flowering lime tree saw, but I feel keenly that it disturbs long tranquil fibers, that it rouses from deep slumber memories linked with beautiful days; I found a veil between my heart and my mind that it would perhaps be sweet— perhaps sad—for me to lift." So wrote Ramond in 1789.[95] Obermann told his correspondent that the odor of mown hay at Chessel made him remember the "lovely barn where we used to jump when I was a child."[96]

Yves Castan, following Lucien Febvre and Robert Mandrou, has shown that hearing was for a long time accepted as the sense of social communication, as opposed to sight, source of intellectual knowledge.[97] The increased role of sight was manifested in legal procedure; in the courts, hearsay was gradually subordinated to eyewitness testimony. But from the mid-eighteenth century on, a new aesthetic movement tended to make olfaction the sense that generated the great movements of the soul.

"Odor," noted Saint-Lambert, "gives us the most intimate sensations, a more immediate pleasure, more independent of the mind, than the sense of sight; we get profound enjoyment from an agreeable odor at the first moment of its impression; the pleasure of sight belongs more to reflections, to the desires aroused by the objects perceived, to the hopes they give birth to."[98]

The sense of smell, by virtue of the very transience of its impression (sad tribute to the power of odor to penetrate consciousness), aroused the sensitive soul, which was then unable to flee from the sensations it had inspired. A strange correspondence was established between the transience of indescribable odor and the revelation of that vague desire, without hope of satisfaction, that was the basis of narcissism. "Jonquil! violet! tuberose! You last but moments!" regretted Obermann, fascinated yet disappointed by the precariousness of the feelings they suggested.[99] Of all the senses, smell was best able to produce the experience of the existence of an "I" conceived as the "contraction of the whole self around one single point."[100] The means of access that it offered to the inner void differed from the approach offered by listening to the rhythms of water. Given his importance in defining new notions of the self, Rousseau's flagrant anosmia, at least in his writing, has probably led later historians to minimize the importance of the role of the sense of smell in the period.[101]

It was already apparent that the sense of smell reveals more idiosyncrasies than does any other sense. "Everyone has his unknown nervous disposition," noted the author of the article "Odorat" in the *Encyclopédie,* concerning the close connection between the respiration of odors and the discharge or retention of vapors. The intolerance to musk henceforth evinced by the elite reflected the intensification of individual sensibilities. Disorders, which we are tempted to see as allergic, were analyzed in terms of idiosyncrasy for nearly a century. According to the most prominent osphresiologists, notably Hippolyte Cloquet, behavior in relation to smells gave vent to the individual's most secret inclinations and imposed them on the whole organism.[102]

The inner experience resulting from the transient effect produced by the odor of the ephemeral flower should be compared with the reaction to excremental odors. We have seen how obsessive perceptions of the progress of putrefaction in living bodies proved at that time. "We live in infection, carrying an intolerable odor inside ourselves always," exclaimed Caraccioli.[103] The place for defecation became specific and individualized. The privatization of waste tended to make it the place for an inner monologue. The only English type water closets installed at Versailles were reserved for the king and Marie Antoinette.[104] Thus they were among the first individuals in France to experience a new privacy. Their example favored that individuation of social practices which encouraged narcissism. Tombs soon became individual and lost their stench. There was already a tendency to discharge hospital patients who could not be cured because they themselves were putrefying. In 1813 Fodéré recommended the exclusion of the scrofulous, "constantly confined in an atmosphere impregnated with the putrid emanations from their bodies."[105]

The sense of smell, again more than the other senses, permitted the individual to experience the harmony of the organization of the universe. Natural odor, by its very transience, introduced into feeling something of that universal harmony which rendered death incomprehensible and allowed hope of a better world. The "shock of the transient" became a "sudden call."[106] Robert Mauzi's analysis throws considerable light on the depth of the change: "Unity between nature and man gives the illusion of a unity within man. Sensation reestablishes the thread that was broken between heart and mind. A simple perfume grows into a sudden awareness of self. This has the effect of associating the 'I' with nature, until then foreign to it."[107]

Experience of this coexistence encouraged a new sensuality that was no longer instinctive voracity but, so Watelet defined it, the art of the "most perfect relationships among external objects, the senses, and the state of the soul."[108] No one expressed better than Senancour that vertiginous power of the sense of smell available to individuals endowed with refined sensitivity. Spring flowers issued to the chosen soul sudden calls to "the most inward life." "A jonquil was blooming. This is the strongest expression of desire: it was the first perfume of the year. I felt all the happiness set aside for man."[109] "Most people could not conceive the relationships between the odor that a plant exhales and the way to happiness in the world. Ought they to regard the sense of these relationships as an error of imagination because of this? Are these two perceptions, which to many minds seem so foreign one to each other, foreign to the spirit who can follow the chain that joins them?"[110]

It is impossible to overemphasize the importance of the flower of the field, with its shy, natural, capricious perfume, a free gift, an infinitesimal ripple that taught the true value of the first stirring of the heart.[111] Revealing unfathomable desires, it was the model on which the image of the young girl was structured.

At the end of the eighteenth century gardens and mountains became places for a multiple quest.[112] The traveler came not only to seek repose or sensual pleasure in their perfumed solitude. Flight from the putrid crowd allowed for the possibility of reminiscence, awakened narcissism, admitted a presentiment of universal harmony, favored the outpourings of the solitary lover. But it was the odor of the jonquil, more than the sight of the landscape composed by the garden artist or the contemplation of rocky vastness, that led to this new sensuality. Later, the functions attributed to springtime scents would progressively devolve on perfumes during the era of the aesthetics of smell.

But for the time being, the focus was on the deodorization of bodies and the environment in order to produce the sensory calm deemed indispensable to the voluptuous disturbances of the "I." Medical instructions for checking putrid fevers and stemming miasmas, the metaphysical anxiety engendered by the advance of putrefaction to the very depths of being, the rise of narcissism and the desire for physical access to scents that it aroused, the wish to be perpetually on the alert to receive natural odors revealing the existence of the "I" and the harmony of the world, the fear aroused by social emanations that were still confused and undifferentiated—all these factors combined to promote the deodorization tactics put into effect from the mid-eighteenth century on. These facts explain the lowering of the threshold of tolerance for stench, the emergence of a fashion for delicate perfumes, and the limited advance in bodily hygiene. Moreover, beyond its formal articulation in medical discourse, this revolution in perception spread, polymorphously, throughout the whole of society.

PART TWO

Purifying
Public Space

6 The Tactics of Deodorization

THE INCREASED INTEREST in public health at the end of the eighteenth century has been the object of a large body of work. My purpose here is not to attempt a comprehensive review of this body of work, but to engage in a rereading of its forms of discourse from the point of view of a history of sensory experience. Public health policy of the period not only borrowed from an already long tradition regarding noxious smells—practices inherited from ancient science that had reappeared in urban regulations in about the fourteenth century; it also took advantage of recent advances in both medicine and chemistry.

The new public health tactics were no longer implemented on an ad hoc basis, as was the case when epidemic had raged; they were meant to be permanent, and they coordinated decisions from within an essentially civic perspective. "The invention of the urban question," the triumph of the functional concept of the "town-machine," stimulated a "topographical toilette" inseparable from the "social toilette" of cleaning streets and organizing places of confinement.[1] From the 1740s on, a form of sanitary policy grew up whose main aim was to develop a fully coherent strategy. It was a strategy directed essentially at doctors whose immense prestige was seen to derive, if not from their practical achievements, then from the authority conferred by that body of "transparent knowledge" deemed to be "indifferent to private interests." The urgency of a plan for social well-being was also emphasized by the nascent science of demography, which, in identifying the town as a site of death, reinforced notions of urban pessimism.

Disinfection—and therefore deodorization—also formed part of a utopian plan to conceal the evidence of organic time, to repress all the irrefutable prophetic markers of death: excrement, the product of menstruation, the corruption of carcasses, and the stench of corpses.[2] Absence of odor not only stripped miasma of its terrors; it denied the passing of life and the succession of generations; it was an aid to enduring the endlessly repeated agony of death.

Paving, Draining, Ventilating

The most traditional feature of this deodorizing hygiene was the effort to isolate aerial space from exhalations from the earth. Constant concerns here were to cut off the flow of subterranean blasts, gain protection from upward currents, stop the impregnation of the soil as a guarantee for the future, and, as far as possible, contain stench. Wherever drainage proved impossible, it was important to cover slime, submerge terrible fissures, and prevent the escape of the effluvia. When the basin of a port or even a canal subject to tidal flow had to be dragged, it was better to wait until the waters had covered it over again.[3] Chaptal advised sanding up the edges of swamps.[4]

The same concern explains the anxious attention given to the "mysterious art of paving," minutely codified by Abbé Bertholon.[5] The classical idea of the city meant that towns maintained their tradition of paving roads, in imitation of the Romans. Paving was pleasing to look at; it made traffic easier; it could be washed by being flushed down with water. But above all, it was a means of sealing off the filth of the soil or the noisomeness of underground water. In sheds adjoining markets, paving stones were indispensable.[6] Caen, which was particularly threatened by stagnating water, undertook paving recurrently.[7] The new pavements, large granite flagstones rather than cobbles, imported from England and only slowly adopted in France, had originated from the same requirement. The first such pavements laid in Paris, in 1782, edged the rue du Théatre Français (the modern-day rue de l'Odéon).

90 ·

Contemporary experts advocated extending paving to village streets and to the interiors of peasant houses. Howard advised replacing cobblestones in hospital courtyards by flagstones.[8] Surprisingly detailed directions were given for paving cesspools, the only way of damming up effluvia.[9] And yet, noted Alfred Franklin, paving introduced a dilemma.[10] It might hamper the upward movement of stenches, but it also stopped infiltration and delayed the cleansing of the earth by rain; it prevented the renewal of underground water and therefore the elimination of past infection. In short, it encouraged stagnation.

In the service of their own ends, sanitary reformers revived the Old Testament injunctions against the moldy infections of houses. Taking down old plaster and replacing it, piercing walls, removing bricks in direct contact with the earth because they absorbed the putrid substances mixed in it were not just sound recommendations for construction.[11] Plastering, coating, painting, and whitewashing walls, vaults, and woodwork provided a positive armor against miasma. Plaster was deemed not only pleasing to the eye but also an effective agent in the struggle against infection. Banau intended his antimephitic varnish to be used for walls and furniture as well as for clothing.[12] Howard expressed delight that the glazed tiles at the Corte hospital covered the cloisters to a height of eight feet.[13]

The desire to seal reservoirs of stenches hermetically seems quite ordinary. Yet it was this desire that governed the tactics used by later sanitary reformers against pollution by the smells of industry.[14] Perfecting the art of designing expertly insulated structures would then justify the presence of chemical factories in the heart of the town. This sealing-off procedure was first used in rudimentary form for excrement. Abbé Bertholon demanded that cesspool-clearing carts be well constructed; he suggested models. Thouret was pleased to note that the majority of these conveyances were henceforth sealed off with plaster.[15]

Despite the importance accorded the circulation of liquid masses, use of water remained ambiguous. Cleaning did not mean washing so much as *draining;* the primary goal was to ensure the discharge, the evacuation of rubbish. From an organicist viewpoint Harvey's discovery and his model of the circulation of the blood created the requirement that air, water, and products also be kept in a state of movement. Movement was salubrious; "nothing can actually become corrupted," Bruno Fortier has noted, "that is mobile and forms a mass."[16] Physiocratic doctrine transposed this rule to the economic sphere. Recognition of the functions of circulation, Jean-Claude Perrot has emphasized, announced a change in urban perceptions; it accelerated drainage trenches and the "knocking down of fortifications."[17] The therapeutic efficacy of movement encouraged canalization and the expulsion of rubbish; it justified the importance attached to slopes. Drying out the town by drains lessened the danger of putrid stagnation, ensured the city's future, and in artificially crowded places afforded the technical control that nature alone could not achieve.

The idea of draining of pestilential swamps situated near towns was widely discussed. Voltaire suggested this method to make the neighborhood of Ferney healthy.[18] In 1781 the marquis de Voyer condemned the marshes that surrounded Rochefort. Bernardin de Saint-Pierre publicized the virtues of drainage.[19]

More important in terms of urban salubrity was the drainage of roads. Cleaning carriageways was a long-standing preoccupation. Jean-Noël Biraben has noted that it was already evident at the time of the Black Death, in the fourteenth century, particularly at Narbonne.[20] Over time, tactics improved. In 1665–66 fear of epidemic was the motive for cleaning the streets of Amiens; the authorities ordered that mud and rubbish liable to spread "bad air" be removed.[21] When an epidemic actually struck in 1669, health measures to combat the infection multiplied; orders went out to slaughter cattle and poultry and to dig latrines in every house. The situation in Amiens was typical. Pierre Deyon recorded identical practices in the Agenais at that time, as well as in the Ruhr and the Antwerp region.[22]

In the eighteenth century, sanitary cleansing became more specific and almost a daily requirement. Cleaning the streets of Paris was made the subject of a competition in 1779. The problem of constructing sewers was already the subject of constant debate.[23] Plans to contain and evacuate sewage abounded. The privatization of refuse was second only to the privatization of excrement as a source of inspiration to authors. Chauvet advocated taking Lyons as a model. In that city "there are containers on every story of the houses, where sweepings are stored; peasants from the vicinity come and remove them regularly every week."[24] Tournon suggested substituting iron vessels for the stones where rubbish was deposited; he also recommended building small boxes onto each house, level with the façade and the pavement, like a basement window with a sliding door.[25]

The reformers nursed the plan of evacuating both sewage and vagrants, the stenches of rubbish and social infection, all at the same time. Bertholon suggested using beggars to do the sweeping; Chauvet wanted the poor and infirm to fill this role.[26] Berne, Lavoisier noted with admiration, was the most cleanly kept town. Convicts "drag large four-wheeled carts through the streets every morning, chained to a pole; longer, lighter chains bind hardened female criminals to the same cart . . . half of these women sweep the streets, the other half load the rubbish onto the cart."[27] Mathieu Géraud suggested entrusting the task of purifying the city to numbered convicts, shackled to a ball. They "would sweep the streets and load the filth from them onto carts drawn by their comrades. They would also remove the ooze from drains and cesspools, the corpses of large animals, such as horses, mules, etc.; they would take the small ones, such as dogs and cats, together with the ooze into which they are normally thrown."[28] Every day they would take away the barrel containing all the household's refuse and excrement and replace it with the previous day's bin, well washed.

Arlette Farge and Pierre Saddy have analyzed the repetitive texts of the medical ordinances.[29] Drying up the streams running down the center of the roadway by prohibiting overflowing gutters (1764); forbidding the dumping of matter and slops (1780); making sweeping the frontages of houses compulsory; ensuring the watering of promenades, bridges, and quays (1750); having well-closed vehicles remove household rubbish deposited next to boundary markers every morning; reforming techniques of cesspool clearance; generalizing the cesspool system—these were the principal measures designed to map the various stages in the proposed "rubbish cycle."

The wish to revolutionize cesspool clearing was the major element in the new medical policy. The reason is obvious. Ever since the ordinance of November 8, 1729, master cesspool clearers had enjoyed a monopoly. They were, on the other hand, subject to increasingly detailed regulations. The ordinance of May 31, 1726, forbade them to let waste matter run out into the "streams" in the streets or to throw it into the Seine or into wells. The workmen had to take care not to use unsound bins. They were ordered to sweep, wash, and clean the ground after them. They could work only at night. They had to go straight to the refuse dump and avoid calling in at taverns. This catalog of instructions makes it possible both to locate the forms of abuse and to discern the genesis of future systems of control, for which the refuse workers were undoubtedly the test-ing-ground.

In 1777 disinfection of cesspools was made the subject of a competition.[30] Over twenty experts, including some of the most re-nowned (Fourcroy, Guyton de Morveau, Hallé, Lavoisier, Parmentier, Pilâtre de Rozier), participated and tried to discover the best disinfectant by analyzing the mephitic gases.[31] Stemming the stenches would ensure that drainage was innocuous.

Removing rubbish without using water involved providing more refuse dumps, both those intended to receive mud and household waste and those intended to receive excrement and decaying carcasses. Whereas rubbish dumps became more numerous in the capital, the open sewers in the faubourgs St.-Germain and Enfant Jésus (near Vaugirard) were abolished (1781). Thus began the long monopoly of the waste-disposal complex at Montfaucon, whose existence became an obsession.

The policy of using containers instead of sewers, proposed at first as a means of combatting foul odors, proved largely ineffective for the time being, at least in Paris. The only advance of any importance was in cesspool clearing. Otherwise, if contemporary accounts are to be believed, the stench only worsened. The streets of the city had been cleaner twenty years before, Ronesse claimed in 1782.[32] The greater number of carriages, the abolition of projecting gutters that watered the "streams," the use of glass windows in shops, which caused merchants to leave their frontages unswept, explained the larger quantity of dirt. How much this analysis owed to the demands of the new sensitivity remains to be measured.

Ventilation henceforth formed the crux of public health tactics. The flow most important to control was the flow of air. Even more than draining away filth, ensuring the circulation of the aerial fluid was a response to the terror of stagnation associated with the coldness and the silence of the grave.[33] Theory justified neo-Hippocratic aerism. Ventilation restored the elastic and antiseptic property of air.[34] Hales emphasized that atmospheric movement also purified and deodorized water that had been corrupted by stagnation.[35] Finally, ventilating "swept away" the lower strata of the air,[36] "restricting the wild circulation of miasmas,"[37] controlling the morbid flow in places where nature was unable to do so freely. Deodorization was both outcome and proof of this control of currents.

Effective ventilation also authorized the strategy of permanent continual surveillance (so strongly emphasized by Michel Foucault). But from the viewpoint of our study there are more important factors than the linkage between absence of odor and the surveillance of behavior patterns. The history we must tackle here concerns the foundations of the rise of narcissism. Techniques of ventilation, insofar as they acknowledged the need for space between bodies and gave protection against other people's odors,[38] brought individuals into a new encounter with their own bodily smells and, as such, contributed decisively to the development of a new narcissism.

In practice, the use of wind machines (particularly bellows) and artificial suction produced by a source of heat coexisted as means of ventilation. Gauger's *Mécanique du feu,* published in 1713, though of little immediate practical significance rapidly became the reference work. Gauger concentrated on private space in an effort to make the light workload and intellectual leisure of the great more comfortable. His primary aim was to heat and ventilate simultaneously the chateau library, the ladies' chamber, and the sickroom by controlling the flows created around the fireplace. By restoring the elasticity of the air he also hoped to stem female diseases. In 1742 Arbuthnot borrowed the same process, although the correct arrangement of the air was only one component of his project since it concerned only sickrooms.

The second third of the century brought decisive advances. In 1736 Désaguliers, inspired by Teral and by Gauger (whom he translated into English), succeeded in renewing the air in the House of Commons by using a ventilator that looked like a windmill and operated by centrifugal force. The duke of Chandos installed two of these machines in his library; they worked for more than a quarter of a century. In 1739 Samuel Sutton suggested ventilating ships by suction from furnaces installed in the bowels of the vessel.[39] Two years later Hales and the Swede Martin Triewald built mechanical ventilators that used bellows.

Until the end of the century people were content to discuss the respective merits of the different processes and meekly decide in favor of one or the other. In 1741 Triewald's apparatus was successfully used in the Swedish fleet; Hales's functioned in several coal mines, at the Winchester hospital,[40] and in Newgate prison. Hales used a windmill placed on the roof of the buildings; in the absence of wind it was operated by hand or by animals. The resulting breeze was reserved for "innocent prisoners."[41] Sutton tried out his apparatus on two ships at Deptford and Portsmouth; as early as 1741 it was used on several British naval vessels.[42] In France in 1759 the vicomte de Morogues and Duhamel-Dumonceau unsuccessfully proposed installing the new machines on the king's ships.

In fact the only apparatus that was used fairly widely, at least in Paris, was the cesspool ventilator, intended to make clearing odorless. A wooden cabinet containing several pairs of bellows was placed over the opening of the cesspool before the actual clearing began. "Wind is introduced into it by three nozzles, two of them horizontal"; the vapors were expelled at a high altitude, "beyond the range of the senses." Its effectiveness was indisputable. Thanks to the ventilator, its inventor claimed, "clearing the cesspool has become . . . a process that is scarcely noticed in the house where the work is carried out."[43] His claim was confirmed by the members of the commission established in 1778 to observe its effects.

Apart from the use of fans, the most elementary ventilators, a bizarre array of techniques was used to ventilate both public and private places. Some doctors advised shaking sheets violently to renew the air in sickrooms.[44] Ingenhousz suggested opening and closing all the doors in apartments simultaneously to cause drafts.[45] His suggestion, frequently revived, gave rise to polemical dispute. Howard declared his support for the practice and advised extending its use to hospitals.[46] Banau and Turben suggested planting plane trees, poplars, elms, and birches around the edges of swamps. All of these trees had vast branches, and their swaying tops would sweep the lower strata of the atmosphere.[47] To the same end, they advised installing in these putrid places windmills with horizontal rotation blades and putting them on sledges in order to benefit different sectors of the unhealthy region. Baumes preferred bellows or windmills similar to the one installed at Dresden on Forestus's advice.[48] Jean-Baptiste Monfalcon recalled that a Bresse doctor recommended "dancing as an excellent means of neutralizing the deadly effects of swampy emanations."[49]

The circulation of vehicles in towns was the subject of some astonishing analyses. The carriage proved to be a very ambiguous device: it was a refuge from the emanations of the crowd,[50] but it was also itself a crowded place and therefore dangerous to the individuals who used it. This hazard was exacerbated by the facts that its jolting hampered digestion and that excessive use hastened the onset of gout and rheumatism.[51] On the other hand, carriages became ventilators for the town as a whole, and their increase was therefore to be encouraged.[52]

Ringing bells and shooting off cannons were still the major methods of ventilation when the putrid threat became more acute. Navier thought that soldiers had been less healthy in the days when men fought with cold steel; cannons purified and deodorized the air of battlefields corrupted by corpses and carcasses.[53] An unexpected side effect transformed them into agents of salubrity: clearing the atmosphere by explosions disinfected it. Jean-Noël Biraben has pointed out that from the seventeenth century on, aromatic fumigations were strengthened by the addition of sulfur and, often, gunpowder.[54] Baumes envisaged purifying the air of swamps by mining the terrain.[55] Banau and Turben were in favor of setting up batteries superimposed on one another.[56] In 1773 powder was exploded inside the church of Saint-Etienne at Dijon to drive out the stench of corpses.[57]

Efforts to control water flows also involved ventilation. It was assumed that a healthy interplay was set up between air and water. The wind cleansed rivers and ponds. Stirring up the atmosphere of swamps ensured that the water was salubrious, on the analogy with shaking a vase; for to shake the contents of a receptacle was enough to purify them. Waterfalls remained the best bellows; the disturbance in the current was communicated to the atmosphere. Banau and Turben, who took the fantasy of the power of ventilation to its extreme point, recommended installing cascades in the center of ponds to produce jets of water and create a splashing effect. In an extreme variation on the same principle, they advised placing small waterfalls at each end of the dining room table and keeping goldfish in bowls for the sake of the disturbance they communicated to the water.[58]

Watercourses, channels of running water, contributed to the salubriousness of the city. Well managed, they could figure among the most effective controls. One of the most persistent dreams of Parisian sanitary reformers was to confine the Seine between two solid rows of quays. This would force it into a permanent state of redemptive disturbance that would prevent the noxious stagnation of carcasses and refuse. Bruno Fortier has drawn attention to the multiplicity of projects intended to control and mobilize expanses of water.[59] The circulation of the aerial blasts engendered by the riverbed, channeled in this way, deserved as much attention as the size and speed of the currents in the water.

Appliances designed to control and harness the natural movements of the air played a considerably larger role than did mechanical bellows or artificial suction. The only ventilator widely used on ships, even in the English fleet, remained the fan-sail, which made the air rush in between the sides of the vessel. Despite its obvious disadvantages, since it could not function in calm weather and slowed the progress of the ship, the ventilating sail satisfied sailors; for a long time they opposed its replacement. It was also used in certain buildings designed for collective use. Howard mentioned one at Maidstone prison.[60]

Fresh air continued to be regarded as the main form of medical protection. Huts, cabins, and shanties, built outside the town in windswept places easy to disinfect by fire, were used for centuries to restrain the advance of epidemics. Invalids were packed into them.[61] The "wind room," together with the "perfume room," remained an essential component of the lazaretto up to the mid-nineteenth century. Once suspect merchandise had been unpacked, it was left there to be exposed to purifying drafts.

The influence of aerist theories on Enlightenment architecture is well known. Planners aimed "to use nothing but architectural resources to capture the air, cause it to circulate, and expel it."[62] The building must be designed so as to separate putrid exhalations from currents of fresh air, in the same way that fresh water had to be divided from used water. The idea was that the shape of the building itself would ensure satisfactory ventilation, thus rendering traditional methods redundant. Cupola and dome were transformed into machines to draw up miasmas; experts climbed onto rooftops to breathe the invisible, evil-smelling spirals they created. The degree of stench was a measure of the architect's efficiency. The Lyons hospital was taken as a model in this respect.[63] Jacques Germain Soufflot designed a vaulted room with an elliptical shape that eliminated stagnant recesses and set up rising air currents.[64] The purpose of arcade henceforth was to allow lower parts of buildings to be ventilated and upward currents channeled. Porticoes were a source of ventilation and at the same time provided pedestrians with shelter from the vagaries of the air. The enlargement of doors and casements, the system so often recommended of locating openings opposite each other, the widening of corridors, and the criticism of towers and spiral staircases (regarded as suction pipes for stenches) demonstrated the growing aerist obsession.[65] Architects liked hatches, airholes, and fanlights. The need for heating tended to take second place to the need for ventilation. Howard went so far as to condemn glass win-dowpanes, although their use was nonetheless increasing.[66]

This obsession caused cellars, vaults, and secret rooms to be denounced on two fronts: they were subject to emanations from the earth, and they lacked the requisite circulation of air. Caverns in-spired terror. Abandonment of the ground floor in favor of the first story began to be advocated. Baumes thought that the masses gen-erally should be compelled to move up a story.[67] Such beliefs pro-voked new criticism of the rural norms of habitation. The sanitary reformers' advice was heeded, as changes in both living patterns and architecture attest. Jean-Claude Perrot has noted the beginnings of migration to the upper stories in the city of Caen.[68] Newly built apartments were ventilated better than old dwellings. Claude-Nicolas Ledoux praised the steps that gave access to raised buildings; they not only symbolized the grandeur of the construction but also at-tested to the belief in the purifying quality of air.

The arrangement of furniture was reconsidered for the same reasons. Particular attention was paid to the bed. Howard insistently repeated his view that the primary requirement was that it be mov-able. Beds should also be fresh, clean, and some distance from each other. He urged that they be placed in the middle of the room and kept from contact with the ground.[69] Tenon advocated iron beds—wood was subject to impregnation—with an openwork base laced to the bedstead.[70] Later, in the early nineteenth century, hammocks enjoyed a great vogue in penitentiary establishments; they met the need for ventilation and saved working space. Models came from other countries: for example, the very high iron beds placed in the middle of rooms in the Antwerp orphanage.

Utopianism was grafted onto utilitarianism. Prevention of crowd-ing, another major requirement of sanitary reformers, would allow control of individual emanations. Le Roy suggested providing an individual outlet at the head of each hospital bed.[71] The patient, weltering in his own atmosphere, would then be protected from the odor of others, not by a barrier but by the controlled flow of air. The plan was in diametric contrast to the centuries-old design. Il-lustrative of the revolution in contemporary thinking on sanitation, these considerations also inspired the debate on ventilation of pris-oners' cells in the following century.

Town planning during the Enlightenment was governed by identical beliefs. The healthy town, an idea popularized by Abbé Jacquin in 1762, would be built on a hillside; the absence of high walls would allow the wind to "sweep vapors and exhalations away."[72] Trades that caused unpleasant odors (tanning, fulling, dying) would be moved outside the walls, together with cemeteries, hospitals, and butchers' shops. Factories would be located in the suburbs. Wide streets, vast squares dotted with fountains, would facilitate the circulation of air. Géraud, for the same reasons, called for "the walls of our cities to be overthrown."[73] Roads, wrote Baumes, should be raised; the ruins and debris of uninhabited houses could be used for this purpose.[74] The model hospital, elaborated in innumerable plans, was designed as a pavilion, like an "island in the air."[75] Claude-Nicolas Ledoux's ideal town at Chaux encapsulated the aerist trend; the houses and public buildings were "independent of all adherence."[76] The town's obvious functional qualities, the separation of its buildings, and its symmetric design, which also corresponded, at least partially, to a requirement of hygiene, ensured that the town was not only salubrious but also immediately comprehensible and visually pleasing to on-lookers.

The king's declaration of April 10, 1783, demonstrated the wish for concrete achievements. The battle against bad air was engaged at the highest level. Norms were established, especially for width of roads and height of houses, to prevent circulation of the fluid from being hampered. It is difficult to measure to what extent they were implemented. However, Maurice Garden has noted that trafficways in Lyons were widened at that period.[77]

Uncrowding and Disinfecting

Uncrowding people and instituting a new division of the amenities of urban space were deemed effective means of achieving ventilation, controlling the flow of exhalations, and damming up the morbific effect of social emanations.[78] The crowding together of bodies was a constant challenge to natural equilibrium and called for a sanitary administration capable of establishing regulative norms. Those considering the problem of the distribution of space gave an essential role to smell.[79]

The body's spatial requirements were to be determined by measurement of exhalations. And the necessary spacings were to be governed by the forms of sensory intolerance we have already noted.

Conversely, over the next few decades, this creation of distance was to entail increasing specialization; eventually, it was assumed, it would eliminate the confusion of smells that often reigned in both public and private space. The privatization of waste encouraged the containment of excremental odors within strictly limited places. Leaving aside the whole problem of intensification that this notion entailed, the argument was that kitchen odors would gradually cease to mingle with those of more intimate spaces, just as those of the hospital would cease to interact with those of the prison.

Half a century later, Villermé spelled out all the social implications of this new goal, which emphasized—for the time being confusedly—the prodigious dangers of both putrid and licentious promiscuity.[80] Attraction to the tangible, warm, and reassuring presence of other people would be openly anathematized. Howard's criticism of prison boiler rooms foreshadowed criticism of workers' lodgings (discussed later).

According to Georges Vigarello, this creation of distance between bodies first took place in the army, as a consequence of instruction in posture, in group movements, maneuvers, gymnastics.[81] Whatever the case, the battle to reduce overcrowding was fought around the individual bed and the tomb, as Jean-Louis Flandrin has recently stressed.[82] The changing configuration of the bed in the eighteenth century was only one stage in the long process of the privatization of sleeping, which began, according to Philippe Perrot, at the end of the sixteenth century when people again began to wear nightshirts.[83] A minority of individuals, endowed with the new sensitivity, perceived the promiscuity and warmth of the collective bed only in terms of the intolerable exhalations of other people. Having a bed to oneself sooner or later implied exclusive attention to the odors of the "I"; it allowed extended narcissistic reverie, encouraged inner monologue, necessitated the private bedroom. Marcel Proust's childhood awakenings would have been unthinkable without this revolution.

All the experts, from Robert Favre to Jacques Guillerme, from Michel Foucault to Bruno Fortier, have recognized clearly the decisive role of the hospital in defining the new norms. It was there, at that moment, that the individual bed became a piece of territory and was transformed into a spatial unit. In this respect, the hospital theorist Tenon played a crucial role: he used the metabolism to justify the need for reform.[84] Every invalid had to be left free to regulate his own heat; it was therefore important to prevent the creation of an average heat that would result from crowding several people into the same bed. Each patient forced into such promiscuity would suffer harm.

Once again, the Lyons hospital was regarded as a model. In 1780, when Necker was prime minister, the Hôtel-Dieu prescribed individual beds. A decision by the Convention on November 15, 1793, enforced this principle as a logical application of the Declaration of the Rights of Man. The attempts at that time to encourage home care were directed to the same goal; they aroused hope that hospitals might one day disappear.[85]

The demand for individual tombs emerged toward the middle of the eighteenth century.[86] If one grave were reserved for every dead person, cemeteries would smell less. What at the time was only an argument for hygiene rapidly became a requirement of dignity and piety. The idea was accepted as early as the beginning of the next century (and thus more quickly than the principle of the individual bed). Inspired by Maret's theory that morbific rays radiated from corpses, Vicq d'Azyr demanded that graves be separated from one another by at least four feet so that the rays emanating from each would not intermingle.[87]

The idea of uncrowding corpses, initially a purely discursive notion, was translated into practice even before the Revolution. Typical was the large-scale removal of the corpses in the cemetery of the Innocents; Thouret was the self-appointed, fascinated poet of this epic event.[88]

Since pure air was the best antiseptic and since the emanations that rose from bodies and filth were incarnations of the putrid threat, ventilating, draining off refuse, and preventing individuals from crowding together constituted disinfection. This word was just as ambiguous as the term *infection,* which denoted the morbific nature and the stench of vitiated air, the preeminence of one type of contamination, and the disruption of organic equilibrium. Meanwhile other tactics were used—apart from the aromatics discussed earlier—to destroy miasmas and restore its original qualities to the contaminated atmosphere. For example, at the end of the eighteenth century, before Lavoisier's discoveries were recognized, chemists frantically sought the antimephitic that would prove capable of conquering unpleasant odor, the danger of asphyxiation, and the morbid threat all at once.[89] This quest accelerated the promotion of chemical disinfectants and deodorants. The main targets were the threats from excrement and corpses.

On the eve of the discovery of the mechanisms of combustion, confidence in the power of fire to disinfect was still intact. Jean-Noël Biraben has shown that the influence of this ancient Hippocratic belief had even increased from the fourteenth century on. In 1348 a whole district of Bordeaux was burned to purify the town; in the next century, the municipal authorities decided to set fire to several houses in the town of Troyes for the same purpose.[90] The great fires lit in Paris to warm the poor during the winter of 1709 probably got rid of scurvy; at least so it was claimed. For this reason, Navier in 1775 recommended an increased number of bonfires in the capital.[91] On August 2, 1720, during the great plague, the Marseilles authorities, on the advice of the Sicards, father and son, ordered blazes to be lit on the city's ramparts, squares, and streets for three days on end—"a gigantic and useless medical *auto-da-fé*" that created a shortage of wood in the town.[92] When epidemics were over, custom decreed the burning of the huts, cabins, and shanties that had served as refuge for expelled patients. The practice of burning contaminated ships persisted throughout the Revolution.

After Lancisi, every expert on swamps advised lighting numerous fires there, particularly when workers were engaged in draining or extracting silt. Pierre-Toussaint Navier prescribed lighting fires when a corpse was exhumed.[93] Lavoisier himself in 1780 recommended this process as suitable for purifying the air in prisons.[94] Duhamel-Dumonceau foresaw the need to disinfect sailors' kit in drying ovens.[95] In 1788 Thouret advocated manufacturing poudrette by means of desiccation.[96]

Scientists did not attribute the same disinfectant power to water. It was more difficult to prevent water from stagnating; in addition, humidity proved more dangerous than dryness.[97] Lavoisier certainly recommended washing prisons—but cautiously. Nevertheless, after his works had appeared, confidence grew in limewater, the first chemical disinfectant; both Howard and Baumes praised its disinfectant powers, achieved through the combustion of lime. Banau and Turben recommended having large numbers of kilns in swampy areas.[98] The mixture that Marcorel concocted to wash the walls of houses and neutralize their putridity worked wonders in the latrines at Narbonne. Howard sprayed the walls of his room with lime;[99] it figured prominently in the disinfection tactics he suggested.[100]

Laborie, Cadet the younger, and Parmentier stated that lime deodorized the sewage water accumulated in cesspools.[101] According to Monsieur d'Ambourney, secretary to the Rouen Academy, steeping some of the product in liquid matter quadrupled the value of the manure; he added that "the odor of the matter is completely dissipated by this mixture of lime and nothing remains of it except an odor something like honey."[102] Lime also deodorized corpses; it accelerated putrefaction of animal matter and combined with "the principle of air" that escaped from the bodies; it dissolved miasmas and prevented them from rising into the atmosphere; it "curbed deadly emanations."[103] It continued to act until the corpse was destroyed. When an exhumation was carried out at Dunkirk in 1783 the use of lime wash stayed the emanations for a time.[104]

But let us turn to the essential. At the beginning of 1773 it was decided to remove the bodies buried in the vaults of the church of Saint-Etienne at Dijon. The stench was so great that the detonating of saltpeter, fumigations, the burning of aromatics, and the watering of the paving stones with *vinaigre des quatre voleurs* failed to destroy it. Neighboring houses were infected and there was danger of fever. Guyton de Morveau was consulted. On the evening of March 6 he prepared a mixture of six pounds of salt and two liters of concentrated sulfuric acid; then he performed a fumigation with hydrochloric acid. Success was total: "The following morning, after everything had been opened to renew the air, there was no longer a trace of unpleasant odor." Four days later, services resumed. Guyton had discovered a "new means of purifying a mass of infected air" absolutely and in very little time.[105] He ushered in the revolution in smell.

At the end of the year jail fever caused thirty-one deaths in the town penitentiary. Guyton set up his fumigations. The following day, "all putrid odor had disappeared, so much so that a medical student offered to bring a bed and spend the night there."[106] The next year Vicq d'Azyr advised the use of hydrochloric acid to disinfect stables ravaged by epizootics in the south of France.[107]

Nevertheless, Guyton's method of fumigation was not used to any great extent before the Consulate. Until then, odor was regarded as the materialization of miasma and was identified with the morbific threat. Although Guyton remained convinced that it displayed the "sort of assimilative power" that forms compounds with a substance putrid in noxious germs, he considered it had the property of a body that had to be destroyed by chemical transmutation.[108] Deodorization would ensure the appearance of a new body.

For Guyton the goal was no longer to mask but to destroy foul odor; "the chemist . . . regards masked odor as nothing more than the confused product of a mixture of elements that continually tend to separate; whereas the destruction of odor is the result of a combination whereby the foul-smelling body is either decomposed or linked to a base that changes its properties."[109] Lavoisier's discoveries later enabled Guyton to refine his theory; in a more general way, he advocated the use of all oxygenants that accelerated combustion of putrid and miasmic substances.

Apparently Dr. James Carmichael-Smith had not heard about Guyton de Morveau's results when he produced almost identical results from fumigations of nitric acid in 1780. His method also permitted "the destruction of unpleasant odors and improvements of the air."[110] It was used in 1795 on the *Pimen* and the *Revel,* Russian vessels ravaged by epidemic. The following year Carmichael-Smith successfully deodorized the military hospital at Forton.

The Laboratories of the New Tactics

Sanitary reformers concentrated their attention on crowded places, where it was urgent to devise some comprehensive regulatory action. It was in relation to these places that the campaign to deodorize bodies and space was first formulated; half a century later, it moved to peasants' houses and workers' dwellings. Soldiers' tents, ships, hospitals, and prisons became the laboratories for the later deodorization of private space.

Apart from the large role played by military hospitals in this process, it was in armies that the first norms of bodily hygiene were tentatively formulated, particularly influenced by Pringle. In 1779, for example, Colombier demanded that the soldiers change their linen at least once a week, and their stockings twice, in order to stamp out foul-smelling emanations.[111] But we must not exaggerate the importance of this disciplinary enterprise. Ordinances, order books, and texts of regulations are extremely circumspect on this point, and this fact implies a paucity of application. Deserters attempting to justify their action referred neither to the bad hygienic conditions of their quarters nor to excessive hygienic discipline; this silence implies that officers were negligent and soldiers unconstrained.[112]

Acknowledging the urgency of the matter, doctors were convinced that the ship had to be transformed into a model of hygiene. Lind set out to codify its salubrity as early as 1758.[113] In France, the vicomte de Morogues defined maritime hygiene down to the smallest detail. He advised pumping out the water in the bilge frequently in order to diminish its unpleasant odor; he forbade meals between-decks; he ordered a relentless campaign against dirt. Members of the crew must wash and comb their hair; the captain must frequently command the decks to be cleared "in order to give sailors' kit an airing."[114]

Cook's ship was acknowledged to be the paradigm: because "every thing accessible to the view [was] completely exposed to the fresh air, the infection it might contain [was] effectively dissipated and destroyed during the voyage."[115] Cook kept constant watch on cleanliness; hammocks and covers had to be brought up on deck whenever the weather was fair; every package was untied and all its contents exposed to the air so that miasmas evaporated during the journey. He inspected provisions to prevent putrid germs. He ordered the airing of spare sails and any fabrics that might be impregnated. Food-stuffs were placed at the bottom of the hold; "during the voyage, the hatches are firmly caulked down, and every crevise completely closed with pitch."[116] Emanations from cargo were kept strictly separate from emanations from crew. Cook's ship was depicted as the first hygienic city in miniature; it was the antithesis of the ghost ship ravaged by pestilence risen from the bottom of the hold. The men in it were protected from rising upward currents of miasma; air and fire rendered the danger from water harmless.

On land, the hospital, and primarily the military hospital, provided a hygienic model. Michel Foucault and François Béguin have shown clearly that it tended at that time to be formed into a machine to distribute air and expel miasmas.[117] As early as 1767, Boissieu clearly laid down the new tactics: the crowded population of wounded persons in hospitals was dying because of the putridity of the air; "to diminish the quantity of exhalations, hospital chambers and wards must be less full; the greatest care taken to keep away anything that can infect, the greatest cleanliness enforced. It is by renewing the air that pernicious exhalations will be driven out . . . Outlets for the air are to be provided by domes, ceilings to be opened up to the next story, fires in fireplaces and not in stoves, Sutton's machine, Hales's ventilators."[118] To facilitate the influx of outside air, doors and windows should be opened, more ventholes provided, pipes placed around each bed. Finally, fumigations should be carried out.

Twenty years later, Jean-Noël Hallé defined tactics that aimed primarily at deodorization. After repeating Boissieu's advice, the father of public health extolled a systematic battle against stench. "The patients will not have their own clothing; the bed curtains will be linen, the commodes will be cleaned and well closed, and the latrines will be arranged so as to give no odor; the place will be swept frequently, particularly after meals and dressings, water will be splashed around with the greatest caution, and sand used in preference for cleaning floors."[119]

Numerous plans were inspired by these beliefs, especially when the Académie des Sciences canvassed architects in 1787.[120] It was envisaged "turning the entire construction into a structure for ventilation";[121] the radial plan predominated in the drawings. Several buildings met the new requirements, particularly the military hospital at Plymouth and the Greenwich naval hospital for veterans in England.[122] Ventilators were set into the ceilings of Guy's Hospital in Southwark, connected with the fireplaces in the upper story; the latrines in this institution did not exude any odor, because opening the door injected them with water.[123] In France, reformers could refer to military hospitals, Saint-Landry's ward (1748), the hospital at Lyons and the hospital Saint-Louis. In 1786 C. F. Viel had culverts and a range of latrines installed at the Salpêtrière;[124] he had already (1784–1786) built the great sewer at Bicêtre, which emptied—inefficiently—into an apparatus that produced compost.

Deodorizing the invalid involved somatic control and, primarily, the monitoring of excrement. Ventilation was not enough; individual behavior patterns had to be changed. In this way, the hospital tended to become a place of discipline. Regulations became stricter. At Haslar hospital near Gosport dirty linen was forbidden; patients' shirts were to be changed every four days, sheets every fortnight. Caps, pants, and stockings had to be replaced once a week. Men had to be shaved every day; patients were prevented from sleeping in their clothes, from using their possessions as covers, from keeping "bread, butter, or provisions . . . upon the heads of their cradles, or about their beds." Neither patients nor staff could "relieve their natural needs except in places intended for this use." Neither "quarrel nor fight" would be tolerated; smoking was forbidden, as was gaming; attendance at divine service was obligatory. "No-one shall be guilty of blasphemous expressions, unlawful swearing, cursing, drunkenness, uncleanliness, lying." At the Chester general infirmary "every patient on admission, is to change his infectious [clothes] for clean linen."[125]

The aim was uniformity, the destruction of age-old habits, the prohibition of spontaneous behavior, henceforth considered anarchic and dangerous. By these premonitory examples, the hospital became the site for an apprenticeship in individual hygiene; there was barely a thought at the time of advocating its extension to the private space of the masses. Tenon envisaged installing "latrines with pans" at the Hôtel-Dieu in Paris.[126] Hospitalized patients and a few privileged souls were then the only people to enjoy this new machinery of comfort.

Identical plans obsessed prison reformers, but in their case the theory posed a dilemma. How to ensure the circulation of water, air, and refuse in places where the circulation of men had to be constrained? How to check the dangers of stagnation and fixity while ensuring the necessary incarceration? How to reconcile the interaction of air currents with the separation of categories of detainees? Ventilation required more and larger openings; jail necessitated confinement. To resolve the problem, Howard advised substituting grilles for doors, bars for panels. Like treadmills, ventilators or, even better, hand bellows could meet both the need for ventilation and the need for exercise.

Control of excrement in prisons posed a similar problem: to relieve the individual of his excrement without jeopardizing confinement. Lavoisier contemplated deodorizing fecal matter from prisons as early as 1780, before experts found a hygienic solution to this problem in the next century.[127] Lavoisier suggested cutting around the institution a channel into which discharge pipes from the latrines would lead. A powerful current of water, created by opening a sluice every two or three days, would drive the waste matter out of the channel; blowoff pipes, terminating in a rotating chimney pot on the roof, would prevent the foul odors from spreading throughout the buildings.[128]

Behavior could be more strictly controlled here than in hospitals, because of the authority vested in jailers. Prisons, like convents though for different reasons, tended to become favored places for apprenticeship in hygienic practices. The propaedeutic value of bodily cleanliness was added to the value that theorists attributed to work. Wrote Howard concerning the prisoners crowded into a prison ship anchored off Portsmouth, "I could wish that the whole of Saturday were appropriated to cleanliness, viz., bathing, washing and mending their clothes, shaving, cleaning themselves and every part of the ship, and beating and airing their bedding. Thus endeavouring to introduce habits of cleanliness is an object of great importance; as many officers have observed 'that the most cleanly ones are always the most decent and most honest, and the most slovenly and dirty are the most vicious and irregular.' "[129]

The requirement of *propre en ordre*[130] and the apprenticeship in housekeeping disclosed their multiple objectives. The aims of moralizing and of repressing the instincts were creeping into areas where hitherto only disinfection had been envisaged. The stench of the sinner was taken literally. The delinquent's rehabilitation was accelerated by his learning how to wash his body. The rebirth of the reformed criminal, ready to receive a new social baptism, should be apparent from his loss of the stench that had linked him with his accomplices.

Dutch prisons were taken as models of personal hygiene. Each inmate had his own room, his own wooden bed, his own palliasse. English prison regulations showed the same concerns. Article VII of the Lancaster prison regulations stated: "The gaoler shall provide coals, soap, vinegar, blankets, straw, mops, sand, brushes, besoms, pails, washing bowls, towels, wiskets and coal boxes for the necessary use of the prisoners, so that their persons and all parts of the prison may be preserved (as much as possible) in a state of cleanliness and health." On arrival, the future prisoner-housekeeper would be stripped of his clothes, washed, and redressed in a uniform. As a preventive measure against jail fever, he had first to be deodorized. Article XII stated: "The gaoler shall take especial care, that every dayroom, nightroom or cell shall be swept clean by one or more prisoners in rotation, every day before breakfast, and washed every Tuesday, Thursday and Saturday." The housekeeping schedule was established. "All such as shall not have their faces and their hands clean washed and their persons clean and neat" would be deprived of rations (article XIII). An extra, better helping would be distributed to the best-kept prisoners on Sundays as a means of encouraging "industry, cleanliness, and good order, and a due attendance on religious worship."[131] There were also a few well-cleaned prisons on the Continent—in Breslau, for example, and, more successfully, the Capitol in Rome.

Lavoisier, who was an advocate of one bed per person, also anticipated making prisoners wash and even bathe when they entered institutions. One of his innovations marked an epoch in the history of ventilation: he recommended that each cell be equipped with two openings, one above the partition to discharge mephitized air, which had become lighter; the other, level with the door to permit the atmosphere to be renewed.[132]

These public health models devised for patients and prisoners are similar in many ways to the model that Vicq d'Azyr tried to impose on stock breeders; salubrious, deodorized stables and healthy, clean, and orderly cattle were also part of the effort to control behavior while preserving collective health.

7 Odors and the Physiology of the Social Order

AT THE END of the eighteenth century, the plan formulated earlier by Ramazzini to write a natural history of odors no longer looked like an impossible project. After the fall of the monarchy, sensualist philosophy had almost official status. Philosophers formed the Section for Analysis of Sensations and Ideas inside the reorganized Institut as early as 1803. However, building up a body of osphresiological knowledge meant formulating a scientific vocabulary. Even from Condillac's point of view, creating a language capable of translating the perceptions of smell involved attempting to separate the sense of smell from the animal origins to which it seemed inextricably linked. Besides, how could the disturbing tangle of the sensations of smell ever successfully be made the object of a discipline without a language that would build them into a system?[1]

There were therefore numerous attempts to define and classify odors. It was a new but tedious venture, steeped in subjectivity, which in the end left scientists with a sense of frustration. Linnaeus, Haller, Lorry, and Virey in turn suggested lists of aromatic categories; none proved exhaustive. It rapidly became apparent that olfactory sensations could not be contained within the meshes of scientific language.

At least the scientists had gained one conviction: the old theory of aromatics was based on an error in analysis. Articles by Romieu (1756) and then by Pierre Prévost (1797) about the gyratory pathways of fragrant particles had already dealt the old dogma a severe blow. In 1798 Fourcroy stated that every type of odor was "solely produced by the simple solution of the odor corpuscle in air or in a liquid." Claude Louis Berthollet later proved this decisively. Thenceforth it was conceded that every substance had a specific odor "related to its volatility and solubility."[2] Theophrastus' ancient statement was transformed into scientific fact.[3]

The triumph of Fourcroy's theory complicated the psychological effects of Lavoisier's discoveries. Understanding of the respiratory phenomena, likened to those of combustion, tended to augment the horror of suffocation, now that its mechanism was understood; moreover, the defeat of the *spiritus rector* theory[4] revived fear of infection and justified continued vigilance in relation to smell. After all, was anything more like miasma than the corpuscle of smell?

For a quarter of a century no one questioned Fourcroy's and Berthollet's theories; Hippolyte Cloquet adopted them. In 1821 Pierre Jean Robiquet posed the problem in a new way: in order to propagate, odor corpuscles had to form a gaseous compound. To do this, they needed a vehicle, an "intermediary." It could be sulfur or, more probably, ammonia. This increased importance attached to the role of ammonia, acknowledged by Parent-Duchâtelet and others, sharpened anxieties about excremental odors.

Thus a scientific osphresiology was laboriously formulated, starting with Linnaeus's books. As early as 1812 Virey drew up a provisional balance sheet and compared recent discoveries with the scientific data of antiquity. In the same year the British scientist William Prout showed that it was really the sense of smell that enabled flavors to be analyzed; on the Continent, Chevreul confirmed this. Finally, in 1821 Cloquet published his impressive *Osphrésiologie ou traité des odeurs,* which remained the reference work until the beginning of the twentieth century. This enormous, sometimes excessive, and much-pillaged compilation contained scientific discoveries, premonitory intuitions, and the most outrageous tittle-tattle side by side. Henceforth writers of manuals and compilers of dictionaries copied freely from it, particularly on subjects related to the hygienic education of the sense of smell.

When Cloquet's book appeared, sensualism was seriously threatened. Clearly the revolution caused by Lavoisier favored physico-chemical analyses at the expense of sensory impressions. Scientists widened their quest in two directions. Some used instruments to track down elusive miasmas, to scan meticulously the alarming scale of filth drawn up in the preceding century. Berthollet analyzed the gases from putrefaction. Chemists drew up precise inventories of gases emanating from cesspools. Boussingault and several others used curious apparatus to try to condense emanations from swamps and analyze the "puterine" they collected in vast linen filters. Chaussier analyzed the products of human respiration. More ambitious, Brachet set out to detect the chemical composition of the subtle perspiration that defined individual odor.

Other experts, armed with eudiometers, attempted to refine the earlier efforts by Abbé Fontana and Priestley to analyze air in public places. Lavoisier, who was the first, obtained significant results. Air "contained in enclosed areas where a large number of people have stayed for a fairly long time" had an abnormally strong carbonic gas content.[5] In 1804 Alexander von Humboldt and Joseph Louis Gay-Lussac detected a decreased oxygen content. On the other hand, following repeated failures by François Magendie, chemists abandoned hope of purifying the air in towns when they were unable to locate any difference in the composition of the atmosphere in the different districts of Paris. As Forget noted, "the triumph of purifying agents is [consequently] restricted to limited areas."[6] Research was given a new impetus in the 1830s when Jean Baptiste Dumas and Boussingault perfected a new method of analysis. In particular it allowed Leblanc and Péclet to define the norms governing the salubriousness of space in relation to the carbonic gas content of the air.

However, it would be an exaggeration to claim that the role of sensory experience was eliminated from analysis. Now that it was known that agitation was not the same thing as purification, there was less reference to tactile perception of the passage or flow of air. Once scientists had confirmed that stench did not necessarily denote the vitiation of air, the role of olfaction itself was challenged. However, it was the sense of smell that in everyday practice continued to detect the quality of the fluid. Moreover, belief in the scientifically demonstrable existence of miasma persisted: it was "a substance added to air" that retained all its mystery. "The dangerous thing . . . chemistry has not taught us about; but our senses are more discerning than chemistry; they clearly demonstrate to us the presence of noxious putrid matter in air where men have stayed for a long period." It was still important to regulate behavior on the basis of sensation, and to continue efforts to renew the air "so long as the sense of smell, which is an excellent indicator in these circumstances, still finds some odor in a place where it previously existed in abundance."[7] Even Leblanc persisted in thinking that a "repulsive odor" signaled the presence of miasma.[8]

Contemporary books about analyses of vitiated air and methods of measuring its restoration reveal scientists' disappointment at the imprecision of their instruments and their vexed resort to perception. It was the sense of smell, Grassi acknowledged, that in the end was to be the acid test of ventilation in ships' holds in the same way that it was from the sense of smell that the air in the prisoner's cell was best tested.[9]

Utilitarianism and the Odors of Public Space

The issue of public health became more focused from the Consulate on. The ideologues' ideas in this sphere, and particularly Cabanis's desire that doctors guide the physiology of the social order, found an echo in the ruling classes. Nevertheless some aspects of public health tactics remained tied to the past. Until about the middle of the century, anxiety about smells continued to govern the fight against the threat from rubbish; we have seen why. The majority of complaints by Parisians to the Conseil de Salubrité still concerned the proximity of putrefied animal substances. The experts themselves, despite the optimism they evinced in regard to harmful practices, kept up their diatribe against putrid workshops. The skepticism of people such as Parent-Duchâtelet remained the exception—as evidenced by the stern attacks directed at him.

Increased crowding in the center of Paris created the obsession with "the rising tide of excrement and rubbish."[10] This phantasm took the place of the image of the city-swamp, which had haunted Louis-Sébastien Mercier. The obvious turning point was 1826, when the threat that Paris would be choked by rubbish became explicit. The Amelot sewer was blocked; the sewers of la Roquette and Chemin Vert began to be obstructed; an evil-smelling pool was growing in the very heart of the city; refuse dumps were infecting the city gates.[11]

Like the removal of the dead in the recent past, the "transfer of sewage" was acknowledged to be a priority.[12] The time had come to control the collective physiology of the city's excretion by organizing the regular elimination of refuse. The ragpicker, whose image pervaded Romantic literature,[13] was assigned an essential role: to sort and order domestic rubbish from houses; to collect organic remains, the bones and corpses of small animals, and so complete the work of the cesspool clearers—which had already been supervised to a considerable degree in the past.[14]

The new concern focused not so much on the danger from the excrement piled up and growing old in the cesspools as on congestion, that is, the inadequate circulation of rubbish through the aerial and underground channels intended to eliminate it. During the Restoration this concern was accompanied by growing awareness of the miasmic threat to Paris from suburbs infected by town refuse dumps and by their own waste. The authorities henceforth concentrated on the recirculation of excrement, which isolated experts had denounced since the end of the eighteenth century. Giving official sanction to sentiments voiced by Mercier not long before, the chairman of the Conseil de Salubrité wrote in 1827: "Go out of Paris today and choose whichever road you will, you will not fail to meet a fair number of rubbish carts, and at any moment find yourself to leeward of a real rubbish dump. The approaches to the capital are already and *from all sides* heralded by the putrid vapors breathed there . . . Soon the sense of smell gives notice that you are approaching the first city in the world, before your eyes could see the tips of its monuments."[15] In that year the council proposed edging the outer boulevards with a vast paved ditch in order to drain into the Seine the foul-smelling water that rushed down toward the center of the capital.[16] In 1828 the chairman of the council echoed Thouret's earlier warning: "the soil surrounding Paris is impregnated by this malodorous muck to a great distance";[17] it was, he added, necessary to avoid surrounding the town with its own filth. The outbreak of cholera morbus in 1832 increased this obsession. In 1835 the council's experts decided to visit the rubbish dumps in suburban communes. At Gennevilliers they noted foul-smelling dumps everywhere, along paths and in courtyards.

What was new was the way in which anxiety about filth and waste conflicted or combined with prevailing utilitarianism. Fear of miasma was coupled with obsession with loss. Henceforth attention was directed to the usefulness of rubbish. The desire to recycle waste, in turn, stimulated olfactory vigilance. Contrary to attitudes in the last years of the eighteenth century, most discussion on excrement concerned its profitability. Malodorous effluvia meant both miasma and deficit. Mille gave a somewhat simplistic summary: "Every unpleasant odor signals a blow to public health in the towns, and a loss of manure in the countryside." Stench confirmed the established fact "of loss, the dispersion of elements."[18] The repulsive odor of excrement was evidence of wastage in the same way that the delicate scents of perfume proclaimed irretrievable and unprofitable expenditure. Utilitarianism and the need for economy strengthened the concern with salubriousness: all three ordained deodorization.

The desire for saving prompted innumerable calculations. Political economy evaluated excrement in terms of profit or loss. At the beginning of the nineteenth century a commission of the Institut rejected the idea of throwing all the refuse of Paris into the Seine—not because it was afraid of spoiling the purity of the water, but because it wanted to avoid waste.[19] Parent-Duchâtelet regarded the export of excrement as one of the capital's great potential resources. As early as 1833 he considered dispatching these products by rail. He would have liked the administration to protect the transport companies; he launched an appeal to individuals: "Help them with your money; take shares in these ventures!"[20] In fact, in 1834 Paris produced 102,800 cubic meters of material, and the refuse dump at Montfaucon alone brought in half a million francs every year.

Bertherand estimated that the refuse industry in the town of Lille had a turnover of 30,000 francs.[21] Sponi thought that the English lost 250,000 francs every year by adopting flushing systems and mains drainage;[22] in 1857 an author in the *Journal de Chimie Médicale* calculated that throwing 332,000 cubic meters of sewage water into the Seine was tantamount to losing 275,600 tons of manure. And these are only isolated examples. They tend to strengthen the link that psychoanalysts make between money and stercus. Perhaps historians of statistics and economic science should also take account of that link. The phantasm of loss, which can be traced from Malthus to Pierre Leroux,[23] the desire to ensure the smooth running of the social physiology of excretion, the concern to keep a record of men and goods and to ensure their circulation, were all part of the same process; repress this dimension of the past and it can be only partially understood.

The deodorization of public space, which commanded more attention than ever, henceforth proceeded via the recovery, exploitation, and utilization of refuse.[24] The proponents of this little-known aspect of utilitarianism translated the project of physical retrieval into the register of social representations. All advocated the utilization of social waste in the collection and treatment of rubbish. They calculated the profitability of people on the scrap heap of society strictly in terms of the value of refuse. Thus the vague plans sketched by sanitary reformers at the end of the eighteenth century became a subject of erudite calculation. Prisoners and beggars were now scarcely spared a thought; sanitary reformers now concentrated their attention on the poor and, above all, the old. By collecting rubbish, these groups could repay in part the community's expenditures on them. Rubbish would be involved in the charitable process; it would lighten the burden on the well-to-do. Belgian towns replaced Berne as a model. In Bruges the poor and the old removed the rubbish; the municipality provided wheelbarrows for those who could not afford to buy them.[25] Ghent and Liège successfully implemented the same policy.[26]

In 1849 Chevallier advocated building free public latrines in Paris, for both men and women, and having them supervised by the poor. He recommended that relief committees appoint road menders and street sweepers. The mayor of Stains tried the experiment; he entrusted public sweeping to individuals registered with relief committees. Beginning in 1832 Chevallier repeatedly advised digging a trench to receive sewage on carefully chosen land outside the tips in every provincial town and every rural commune. Once that was done, a few paupers, paid by the inhabitants, "would be given a small cart, drawn by an ass or a poor horse, and ordered to go through the commune and its approaches *without stopping* on all working days, removing all the rubbish there with a shovel and broom, and take them to the commune's depot. This *constant cleaning* would provide quite a large mass of products, and their continual removal would maintain a pleasant and salubrious cleanliness."[27] The nature of the work was the same as for dung collection; but the suggested rhythm was new. The ceaselessly repeated nature of the activity ensured both complete saving and absolute cleanliness, deodorization, and salubriousness.

The collection of human excrement commanded all the more attention insofar as its superiority as a fertilizer seemed undisputed at that time.[28] Liquid or solid, it was the richest fertilizer. "Every kilogram of urine is equivalent to a kilogram of wheat," declared Sponi.[29] Dominique Laporte cites a body of important texts aimed at drawing the prefects' attention to this exceptional property.[30] The ordinance of December 31, 1720, had at one time controlled and encouraged use of these manures in the Paris region. They clearly became unpopular between 1760 and 1780. At the end of the ancien régime, the volume of excrement used fell sharply, except in regions such as Flanders, where its use was traditional. A new period of increasing use began later, corresponding with the rise of utilitarianism.[31] This revival of interest forced the Conseil de Salubrité to define a policy. The desire to avoid confusion of products, and thereby loss, led it in 1835 to advocate the use of apparatus that would separate liquid from solid material.[32]

The problem of mains drainage, suggested by England's example, remained. Not until the last years of the nineteenth century did this solution gain acceptance in Paris, although it had proponents as early as the Restoration. Moreover, the system operated in some parts of the capital: the Ecole Militaire, the Invalides, Bicêtre, the Salpêtrière, and the Mint evacuated their excrement through sewers that discharged into the Seine. Liquids from Montfaucon were collected and conveyed to the river through the large drain that encircled it; then, beginning in 1825, through the drain parallel to the canal Saint-Martin.

This method of drainage was used in many provincial towns until the end of the century, even until World War II. In 1860 the basse Deule was in theory the sewer for the town of Lille; pipes from public latrines and offal from slaughterhouses met there. Unspeakable filth accumulated in the culverts, and the stench invaded the whole town.[33] At Caen the Odons were transformed into virtual open-air sewers and remained so for more than a hundred years.[34] In 1876 the Nièvre at Nevers was nothing but an "immense cesspool."[35] No systematic disinfection of urban space was undertaken here until the end of the century.

Again and again proponents of mains drainage, from Sponi to Guéneau de Mussy, repeated that it was the only way to ensure the movement, the transport of excrement and thereby defuse the terrible threat of stagnation. In addition, unlike cesspools, it permitted control of the flow: "drains are under constant supervision. This is brilliant, easy, regular," Emile Trélat still pleaded in 1882.[36]

Why, then, was this solution rejected for nearly a century? Gérard Jacquemet has shown the complexity of the discussion and the conjunction of interests that acted against its adoption.[37] The system would impose a payment on property owners that was long considered too heavy a burden. In 1856 only 10,000 of the 32,000 premises in the capital were supplied with water. There was a risk that mains drainage would ruin the cesspool-clearing companies, which formed an effective pressure group. These obstacles might have been overcome if the outcry from experts had not strengthened resistance. Their arguments reveal the recurring obsession with loss. Chevreul pointed out that disinfecting the material weakened it—an effect that had been lost sight of in the concern for public health.[38] Aside from mains drainage—the classic case of waste—flushing the cesspools with water decreased nitrogen content. The cesspool clearers were well aware of this; they valued the excrement of the poor much more than the far too diluted excrement of the rich. Marie François Belgrand drew up a precise social scale of the value of the product. He plotted the topographical distribution of the nitrogen content of the capital's excrement.[39] Thus whereas utilitarianism spurred the deodorization of roads and public space, it slowed the adoption of mains drainage in Paris and numerous French towns.

During the Restoration, human waste became a raw material in industrial chemistry. A factory was set up to produce ammonia near the new refuse dump at Bondy. This project met the requirements of both health and utility. Its success seemed the incarnation of that obsessive plan to eliminate excrement whereby the sanitary reformers of the period came to recommend using means capable of instantly transforming dejecta into an excellent manure in latrines.[40] After that, the chemistry of refuse inspired numerous grandiose projects; in 1844 Garnier dreamed of building a vast industrial complex for the treatment of urine, to be called *ammoniapolis*.[41]

A new era of knackery began in 1825. On the eve of this change, the stench at Montfaucon had reached unprecedented strength. The villagers of Pantin and Romainville breathed this pestilence constantly. Parent-Duchâtelet, technical expert on the spread of odors, had just produced a fine study of foul-smelling effluvia.[42] Fortunately, the topography of the area diverted the effluvia and spared most of Paris, but they infected the Combat Gate and, when the wind blew in certain directions, affected the Marais and the Tuileries. Lachaise had complained bitterly about these effects three years earlier.[43]

In his report on the activities of the Conseil de Salubrité in 1815, Moléon had already posed the problem: how "to convert immediately into material suitable for trade the muscles, blood, fat, bones, and intestines of the ten to twelve thousand horses slaughtered in Paris every year."[44] The elimination of the three thousand animals killed on March 31, 1814—the date of the Allied attack on Paris—had underlined the urgent need for a solution.

As early as 1812 the chemists Payen, the Pluvinet brothers, and Barbier had obtained approval for a process of producing manure by liquefying the fats and compressing the fleshy parts of carcasses. In 1816 Foucques proposed "making different colored soaps and an alkaline liquid from flesh, bones, and intestines from the slaughter of horses."[45] After 1825 the new Payen works set up at Grenelle revolutionized this industry. Processing in isolation and the use of bone black transformed sordid knackery into "a salubrious activity that ensured considerable profits."[46]

Using lime to process tallow also rendered odorless an industry that in the past had evoked incessant complaints. As a result of work by the younger Barruel, ammonia salt was produced from bone remains, carcasses, and purifying water.[47] Demand for raw material was an incentive to collect the animal remains that cluttered the public highway; it helped clean up the town. After the Pluvinet factory was set up at Clichy, "walls or heaps of bones were no longer to be seen in the capital's avenues."[48] A large quantity of animals' blood, which had formerly streamed over the paving and infected the air around slaughterhouses, could now be processed at a drying works that exported its products to sugar refineries in the colonies.[49]

The search for profits led to the deodorization of public space more surely than the obsession with unhealthiness. It worked to eliminate the stench of carcasses, blood, and bones. Although the new chemical industry initially also produced foul-smelling effluvia, the upsurge in production techniques in isolation, the growing use of disinfectants, and increasing legislation on insalubrious establishments counteracted these harmful practices.

In this way the obsession with decomposing animal flesh was gradually defused. In the eighteenth century researchers had analyzed in minute detail the putrefaction rate of meat; the process was now accelerated to the point where transformation appeared instantaneous. The model provided by the knackers' trade suggested a way of abolishing miasma by abolishing stench. Parent-Duchâtelet confessed his admiration for the revolution that was taking place. Every part of the corpse, deodorized and carefully sorted, had a rational use.[50] All that remained to be done was to prohibit clandestine abattoirs, which still existed in the capital, to ensure that the new ones had a monopoly on processing corpses. Several commissions, appointed by the Préfecture, formulated the necessary regulations. Compulsory use of high-draft furnaces that carried emanations "to a great height in the atmosphere" completed the deodorization of the rendering process.[51]

An element of taboo still surrounded discussion of the possible utilization of human corpses. Parent-Duchâtelet, ever the pious man, devoted a long article to the abuse to which human fat was put by students in the Latin Quarter dissecting room; but he could not have violated the taboo himself.[52] Not until 1881 did the engineer Chrétien set out a plan, which he described with unfortunate foresight as a progressive idea. "The aim of all burial," he wrote, "should be to transform inanimate remains into useful products."[53]

The Chloride Revolution and the Control of Flows

Scientific deodorization based on correct analysis of foul-smelling gases now progressed at a lively pace. There was increased everyday use of fumigations on the lines suggested by Guyton[54] and, even more, of Javelle water, produced as early as 1788 in the comte d'Artois's factories.[55] Two discoveries completed the work of the Dijon chemist. The pharmacist Labarraque found a way of arresting the advance of putrefaction by using lime chloride instead of chlorine. The decisive experiment was carried out on August 1, 1823. At 7:30 A.M. that day an exhumation took place; the famous Orfila then had to carry out an autopsy. But there was a terrible stench from the corpse. Labarraque suggested sprinkling lime chloride dissolved in water, and this had "a marvelous effect . . . the foul odor is instantly destroyed." The prefect Delavau was very quick to draw the lesson from the experiment; he ordered "the latrines, urinals, and slop bowls in the capital" to be disinfected with chlorinated water.[56] In 1824 Labarraque wrote *Instruction à l'usage des boyaudiers, contenant le moyen de travailler sans fétidité* (Instructions for the use of gut dressers, containing the method of working without fetidity).

The very death of Louis XVIII confirmed Labarraque's success.

The king's corpse was in such a state of decay that it emitted a terrible odor. The pharmacist had to be called in. He impregnated a sheet with chlorinated water and held it in front of him like a screen, then covered the body with it and sprinkled it for a very long time; in this way he successfully removed the unpleasant odor.[57]

Labarraque's bucket of lime chloride rapidly became indispensable equipment for all the great hygienic ventures. In 1826 it enabled the workers clearing the Amelot sewer to be disinfected.[58] The new liquid deodorized the corpses of the July dead in 1830. The July Revolution marked the definitive triumph of chlorinated water. Dr. Troche sprinkled it on the graves under the Marché des Innocents square and in front of the Louvre.[59] A few days later, using Labarraque's bucket, Parent-Duchâtelet stamped out the stench from the bodies hurriedly piled up in the vaults of Saint-Eustache. When cholera morbus broke out less than two years later, the precious liquid was used in an attempt to disinfect the whole capital. The prefect Gisquet ordered it to be used to clean the stalls of butchers and *charcutiers* and to "neutralize . . . putrid emanations escaping from graves, trenches, terracing work."[60] The flagstones in markets, the pavings stones of roads, and the trenches along boulevards were watered with the substance.

Labarraque's discovery made it possible to eliminate the horrible stench that had previously prevailed in dissecting rooms.[61] For medical students and their teachers this was a daily torture; some lived in constant fear of infection. Dissection, like knackery, was dispersed throughout the lanes of the Latin Quarter; the neighborhood complained of pestilence. The ban on clandestine establishments and the daily washing of tables with chlorinated water in the Faculty's new dissecting room helped to deodorize a whole district of the capital.[62]

The terrible stench of hospitals had still to be checked. Labarraque's famous liquid of chloride of soda proved effective. Experts noted, not without exaggeration, that the substance made it possible "to curb decomposition on the living being." Gangrenous carbuncles, "degenerate venereal ulcers," "the most intense hospital gangrene," cancer itself, could henceforth be "disinfected," that is, deodorized.[63]

The other important discovery was the product prepared by Salmon in 1825. It had long been known that coal dust could disinfect; by calcinating animal matter with earthy substances, the chemist succeeded in preparing an animalized black capable of instantaneously deodorizing all "material in putrid decomposition."[64] The substance produced was a valuable manure. Salmon reconciled sanitary reformers and economists. He made the disgusting poudrette obsolete, although it still offended the noses of the heroes of Balzac's *Un Début dans la vie,* confined in Pierrotin's carriage.[65]

Dozens of processes designed to disinfect excreta followed in quick succession after the mid-eighteenth century. In 1856 Sponi listed all the projects formulated since 1762; he counted no fewer than fifty-seven.[66] For nearly a century the greatest scientists pondered and experimented; every eminent chemist tried his hand at deodorizing excrement.[67] First animal black, then iron sulfate finally provided effective solutions. These products soon dissipated the terror inspired by cesspool clearance. Around the middle of the century La Société Générale des Engrais, often quoted as an example, operated in the open air at Lyons without provoking complaints; "opening up a cesspool inside a shop does not stop customers from coming in to buy."[68] Progress in the capital was slower until the ordinance of December 12, 1849, prescribed disinfection of cesspool clearing with sulfate and zinc chloride. The large number of processes suggested seems to have delayed adoption of the best ones. Odorless public latrines installed in the rue Neuve-Saint-Augustin as early as 1817 were not imitated for several years.

Deodorization of crowded places necessitated the control of flows. It was not enough to allow air to circulate; it also had to be directed if the stagnation and stench in corners were to be totally obliterated. Advances in ventilation achieved by Tredgold in England and by the practitioner J.-P. d'Arcet and the theorist E. Péclet on the Continent, promoted this approach.[69]

"A process," stated the engineer Grouvelle, a disciple of d'Arcet, "is never perfect unless it can be controlled at will." Knowing how to evacuate, but also how to direct and distribute, the air in an area was to be able to control the smell of its atmosphere. Closed circuits seemed to be the means of achieving this sort of project. "There is no good ventilation when it is subject to atmospheric variations, the action of the winds, conditions created by opening and closing doors and windows, independently of the process itself." As Grouvelle noted, d'Arcet understood that "fairly steady, fairly powerful methods had to be adopted so that the chief current of air that was to be set up, prevailed over all the accidental currents without variation and without interruption."[70]

Pearson's hermetic room was taken as a model. The English

doctor strove to produce a pleasant temperature so that consumptives could be looked after at home and thus spared a long stay in southern climes; he imagined that he could do this by closing the fireplace, constructing double doors and windows, and ending up with a sort of air-conditioned greenhouse designed for humans.[71] These conditions involved a revolution in everyday behavior. The housewife in the new kitchen that d'Arcet conceived as early as 1821 would be careful not to open doors and windows. "She would have to overcome an old habit," the expert admitted; "in the old type of building she used to open everything so as not to suffocate, she had to let a great deal of air enter her kitchen to render less harmful both the noxious gases and the smoke that filled it, but in our system . . . there is a steady draft; there is no steam in the room."[72]

D'Arcet praised the use of siphons and industrial production in isolation for the same reasons. The new ventilation accelerated replacement of open hearths by stoves and boilers. Since smooth surfaces made it easier to control flows, it was quite logical to acknowledge the advantages of enamel and varnish, which air and water slid off unhindered. The origins of the "clean and decent" bathrooms of the end of the century lay in this desire (expressed as early as the Restoration) to control drafts.[73]

D'Arcet systematically applied two principles in the area to be ventilated: "artificial suction, on the one hand; on the other, *steady distribution of air*";[74] only the second was an innovation. Furthermore, deodorization, according to d'Arcet, presupposed total combustion; the smokeless furnaces that he constructed and constantly advocated were part of the same pattern.

Prisons continued to be a focus of anxiety. In no other place did the essential circulation of air command such keen attention. Villermé considered it indispensable to delimit the area ventilated. He recommended building an enclosing wall, which would eliminate fear of escape.[75] Deodorizing prisoners' dejecta was the second requirement. Inside prisons, the cell became the laboratory for experiments in deodorization techniques. Members of a commission composed of the most eminent chemists (Dumas, Leblanc, Péclet, Boussingault) used their senses of smell to measure the relationship between the time necessary to obtain total deodorization of a filthy cell and the volume of air introduced. Gathered around a stinking bucket for hours at a time, researchers compiled tables "that fix the basis of ventilation and purification of all solitary confinement prisons."[76] It was not the volume of oxygen necessary for the individual's survival that here governed ventilation, but the strength of the current capable of overcoming the stench of the prisoner's dejecta.

Similar experiments were carried out in schoolrooms infected by pupils' sweat and unclean clothes; they showed that six cubic meters of air per individual per hour were sufficient to make any odor disappear.[77] From this experts deduced that a current of twelve cubic meters per hour would succeed in deodorizing places where adults were crowded. These new norms were a source of inspiration to engineers. At Mazas prison, Grouvelle succeeded in "purifying" twelve hundred cells in this way by using "downward suction" from the soil pipe of the latrines. Duvoir obtained good results in the cells at the Palais de Justice by a quite different method: latrines with siphons.[78]

After 1853 Van Hecke's mechanical ventilator was increasingly perceived as a model for penitentiaries. Ducpétiaux confessed his admiration.[79] The anemometer installed in the solitary confinement prison of Petits-Carmes in Brussels showed that the new apparatus provided forty-eight cubic meters of renewed air per person per hour, whereas the authorities demanded only twenty. A dial with a needle was visible from the gallery and indicated night and day "the real strength of the ventilation; a glance is enough to estimate its various gradations, from zero to eight, which is the maximum figure."[80] Although it did not use artificial suction, the ventilator was the realization of d'Arcet's dream. A constant flow, adjustable, continually measured, ensured the control of ventilation that was indispensable to the elimination of personal odors. The machine's reliability ensured its success. "Throughout the experiment, the needle remained almost invariably between the fourth and fifth degrees on the dial."[81] This type of apparatus was installed at Beaujon hospital in 1856. The following year Van Hecke's ventilator worked wonders on the *Adour,* transporting five hundred convicts from Toulon to Cayenne. It accounted for the fact that the chief surgeon did not have to order a single infirmary confinement during the crossing.[82]

Deodorizing places of confinement by means of ventilation involved discipline concerning defecation; authorities unanimously deplored the absence of such discipline.[83] Well before strict practices were developed in schools or private space,[84] they were introduced in asylums to ensure salubriousness, and sometimes also to recycle excrement.[85] Significant in this respect was Girard de Cailleux's requirement that attendants put inmates into iron harnesses night and day at a stated hour, forcing the patients to relieve themselves at the designated spot. Observation had shown that this training was possible. "On the subject of visiting the latrines, just because he is deprived of reason, the lunatic can be subjected, as far as the fault of uncleanliness is concerned, to a repression that it is not possible to impose on ordinary inhabitants of public institutions."[86]

In the same year, 1858, Edmond Duponchel published a fascinating plan that conveyed more sharply the desire to banish the stench of excrement from collective places and to impose discipline by means of the structure of the equipment itself.[87] To deodorize barracks and hospitals, the author suggested building a tower latrine, described as a "minaret." Its baroque architecture, inspired by the ship's tower and probably by d'Arcet's dovecote, aimed at debarring individuals from any possibility of contamination. No longer would floors or walls be stained; patients or soldiers were held in iron harnesses and set upon a virtually suspended lavatory seat reached by a metal walkway.

D'Arcet, like his colleagues on the Conseil de Salubrité, particularly his friend Parent-Duchâtelet, dreamed of making every industry salubrious. He deodorized several of the most foul-smelling factories by installing smokeless furnaces and reverberatory furnaces. Olivier de Serres had already expressed concern about silkworm-rearing houses, "infected by the worms' respiration and transpiration, their excrement, skin changes, corpses, and the fermentation of the litter."[88] In 1835 d'Arcet succeeded in eliminating unpleasant odors from this industry. He "purified" gold and silver refining in the same way, and then the drying of tobacco leaves. As a result of d'Arcet's work, industries with nauseating smells that city dwellers were no longer prepared to tolerate could be set up or continue to operate in town centers.

The mid-nineteenth century saw the expansion of the ventilation industry, its lasting prosperity ensured by long-term resistance to stench.[89] France was clearly behind in this field. Unlike their English colleagues, architects on the Continent were ill informed on advances in physics and scornful of engineers (who were, moreover, few in number); they remained primarily concerned with beauty of form. A visit to Rome enjoyed far greater prestige than did research on the mechanics of heating or ventilation; that was still the province of oddballs. Only rarely did ventilation govern the overall conception of a building, as in the examples mentioned previously. The ventilation theorists were not listened to. What was lacking was a body of civil engineers sufficiently organized to impose itself as an intermediary between architects and mechanics. In England almost all public buildings and a number of dwellings and ships were equipped with ventilation systems, although these were often improvised; in the large towns of the July Monarchy there were only a few exemplary achievements. Sanitary reformers focused their concern on places of public entertainment in Paris, where bourgeois and aristocratic ticketholders spent interminable evenings crowded together. The hall in the Variétés, which d'Arcet ventilated by chandelier suction, was a rapidly and widely imitated model.

8 *Policy and Pollution*

*The Formulation
of the Code
and the Preeminence
of Olfaction*
FROM THE Revolution until Pasteur's discoveries, progress in public health seems again to have occurred through borrowing; legislation drew heavily on the arsenal of measures promulgated during the ancien régime, which has often proved of dubious efficiency. Furthermore, the diatribe against cemeteries, and subsequently against places where the putrid mob crowded, had created a model for anxiety, vigilance, and intervention. Thus in many respects the nineteenth-century debate on industrial pollution was only a culmination of a previous development. What was new was the coherent form of the policy decisions. Beginning with the Consulate, a code was gradually formulated that defined both the harmful practices and the policy toward them. The new public health aimed at increasing the pace of disinfection; its target was all of space and all of society.

The legislative history is clear. Two laws on industrial crafts and salubriousness were promulgated in 1790 and 1791. They had a very limited effect. They did not classify insalubrious establishments; they neither calculated nor defined the damage caused by industry. The courts remained impotent, jurisprudence vague and arbitrary. These legislative measures perpetuated the ancien régime's tradition of ineffectiveness.

The creation of the Conseil de Salubrité of the Department of the Seine on Messidor 18 in year XI (July 7, 1802) provided the administration with a stable organization for consultation and control; it sanctioned new objectives and also a more precise code. At the request of the minister of the interior, the Institut's physical sciences and mathematics division on Frimaire 26 in year XIII (December 17, 1804) proposed a classification of insalubrious and dangerous establishments. This text guided the administration's actions for nearly three years. On February 12, 1806, an ordinance by the prefect Dubois forced manufacturers wanting to start a business to announce its opening beforehand. It also required them to submit a plan of the projected factory or workshop. "Specialists" accompanied by a police inspector were to visit the sites and make an official administrative report.

In 1809, following angry complaints about soda production, the minister of the interior appealed to the Institut again. The 1804 report now seemed too imprecise. For twenty years the proliferation of factories had aroused opinion to the point that careless siting of industry in an urban milieu was not to be tolerated much longer. Slaughterhouses, gut-dressing works, and tallow-melting houses continued to cause alarm. But these were now surpassed in the hierarchy of anxieties by factories making Prussian blue, strong glue, and poudrette, which were said to be springing up in all the large towns in the country. Although scientists condemned the harmful effect of acid vapors much less forcefully than they did putrid miasmas, opinion seems to have been very hostile to factories making vitriol, salt of Saturn, ammonia salt, and, above all, soda, all of which multiplied in the early years of the Empire. Metal gilding and all processes involving lead, copper, and mercury were also considered undesirable in the vicinity of private dwellings.[1]

The emperor's own behavior was evidence of the new intolerance. Disturbed by the foul odor emitted at Saint-Cloud by the waste from the empyreumatic oil factory at Grenelle, Napoleon ordered that henceforth these products were not to be discharged into the river.[2]

The Institut's chemistry division, which was charged with the investigation, asked the chief commissioner of the Paris police to conduct an exhaustive census of industrial establishments in Paris. Consequently it proposed a distribution ratified by the decree of October 15, 1810, which became the reference for all later measures.

The 1815 royal ordinance did nothing more than reproduce its major features.

The main purpose of this legislation was to protect the industrialist against the jealousy or malevolence of the neighborhood, to ensure his peace of mind and thereby permit his enterprise to expand. On their own admission, the Institut's experts intended to introduce industry into the heart of towns, to impose there in the same way that it had earlier been possible to compel toleration of blacksmiths, coppersmiths, coopers, founders, and weavers, "whose profession is to a greater or lesser degree disturbing to the neighborhood." The plan to move workshops to the countryside, nursed during the latter years of the ancien régime, was forgotten.[3] The current tolerance was strengthened by the conviction that the upsurge in chemistry and the advance in "handling fire" would quickly be able to abolish the harmful side effects. Already, according to the 1809 report, some factories making soda and Prussian blue were functioning without any inconvenience.

Anyone at all familiar with the medical, or rather the municipal, literature of the last years of the ancien régime must find its definition of unhealthiness very limited. The alarmist note sounded by chemists at the end of the eighteenth century had disappeared. For a time, optimism governed scientific discourse. Only the presence of noxious miasmas, shown by the deterioration of metals or the withering of vegetation, justified the description "insalubrious." Of course, workshops where "large quantities of animal or vegetable material were accumulated and caused to rot or putrefy formed surroundings harmful to health," but the main trend was to replace the emphasis on the insalubrious with an emphasis on the merely inconvenient. Most chemical vapors did not seem to deserve to be described as unhealthy, since they were "developed by means of fire" and could be condensed. According to the report of 1804, "factories making acid, ammonia salt, Prussian blue, salt of Saturn, white lead, butcheries, starch factories, tanneries, breweries (and even the production of sulfuric acid) in no way form surroundings harmful to health when these factories are well conducted."[4]

Even the idea of inconvenience appeared to be limited to a definition based on smell. Article I of the decree of October 15, 1810, provides clear evidence: "From the publication of the present decree, factories and workshops that spread an insalubrious or disturbing odor cannot be set up without a permit from the administrative authority." The few references to noise occur only as part of an appeal to tolerance. Smoke itself attracted little attention at this stage. Dust was not yet a cause for concern. And a fortiori these texts do not allude to sight; what might shock the eye or diminish light was neglected.

Manufacturers therefore could have felt almost entirely reassured were it not for the landlords, the only effective brake on the uncontrolled expansion of industry. The decisive measure of injury was a fall in the market or rental value of property near a factory. The contest between the two groups was endlessly repeated. On February 9, 1814, the minister of factories lucidly interpreted the sanitation decree of 1810 as simply a conciliation measure between manufacturer and landlord.[5] Workers' health was scarcely considered; neighbors' health remained of secondary concern.

Numerous measures concerning details completed the decree of October 15, 1810. Trébuchet consolidated these texts as early as 1832; they formed a "clear and detailed program for every type of industry and every class of individual."[6] The new legislation divided establishments into three categories[7] and made the system of preliminary authorization universal; it introduced supervision to check the anarchic proliferation of workshops and therefore of hazards and injuries.

The Apprenticeship in Tolerance

The conseils de salubrité, established in the principal towns between 1822 and 1830, existed to ensure enforcement of the new legislation. Engineers, chemists, and doctors sat side by side on them. These experts acted in accordance with the principles that had guided the formulation of the legislative texts. Their conciliatory attitude, however, reveals that they were not the instruments of an authority intent on exercising a meticulous supervision. The councils' function was primarily to reassure, to defuse anxiety aroused by stench, to permit a peaceful life in the vicinity of industry. The optimism they evinced toward harmful practices was based on confidence in the advance of chemistry, in marked contrast to fear of the city's being clogged up by excrement, which continued to haunt them. Inspired by an Augustinian sense of human imperfections, convinced of the need to provide safety valves—in other words, to tolerate necessary evils—the sanitary reformers on the councils worked to promote tolerance. In their eyes, to bring light, to shed light on a hitherto obscure threat or danger was already to defuse it; only light purified; hence they addressed themselves initially to seeking out the clandestine, to dispelling the opaqueness of the potentially menacing. Thus, before intervening they waited till public opinion showed itself through complaints or petitions. They acted more as arbitrators than as inspectors.

This fact explains why the industrial odors that pervaded public space were so slow to disappear, despite the importance legislators accorded to olfaction. A large number of dairy farms prospered within Paris itself with the assent of the Conseil de Salubrité. According to experts, most chemical vapors were not dangerous except to workers who breathed them at very close quarters. This hazard was not enough to warrant closing the establishments. Here again, the idea of unhealthiness was applied only to the neighborhood population, not to the work force. Moreover, inconvenience was not held to be of concern to the work force, since constant exposure ensured that they no longer perceived the hazards and the unpleasantness. "Thus it is," noted the Institut chemists in 1809, "that when one goes into factories producing simple and oxygenized sulfuric, nitric, and hydrochloric acid, for example, one is immediately struck by the odor of these acids, while the workers barely notice them and are disturbed by them only when they unthinkingly breathe in a great deal at one time."[8] "It must be noted," Monfalcon and Polinière elaborated in 1846, "that workers fairly often become acclimated to workshops; very few complain, very few appear to notice the unhealthiness of the environment in which they are condemned to live."[9] It was the statistician's job to measure the dangers of industry to the health of the working population. The workers were insensitive to them and unable to measure them.

By skillful propaedeutics of technical progress, the experts on the councils succeeded in getting neighborhood industry accepted. The process was almost always the same. Initial complaints aroused by innovations were followed by resignation, tacit acceptance. Coal was resisted at the end of the eighteenth century, the subject of vituperation in 1839, and finally accepted; and once coal was tolerated, the steam engine had to be, too. The process was repeated with "distillation of acids," and then with production and combustion of coal gas. Parent-Duchâtelet's behavior exemplified this desire for tolerance which ensured that French towns continued to stink long after the spread of the lowered sensory tolerance.

But the history of the struggle against malodorous unwholesomeness is not only recorded in the legal codes; it is not identical with the triumph of tolerance; it is also made up of ambitious projects and arduous, often sordid, sometimes epic, battles.

After the very limited success of the attempt to clean up the roads in Paris during the Consulate and Empire, the Restoration seemed a time of great ambitions, if not of concrete achievements. The political aims of the sanitary reformers had never been as clearly stated as in this period, which saw the birth (1829) of the *Annales d'Hygiène Publique et de Médecine Légale.*[10] Tactics were formulated for purifying traditionally crowded places, particularly barracks and prisons.[11] The struggle against the rising tide of excrement became a focal point for these efforts. The stench of Bièvre, which had seemed to reach a high point in 1821,[12] was now reduced. The cleansing of the Chemin Vert, la Roquette, and Amilot sewers provided an opportunity for experimenting with ventilation, fumigation, and disinfection techniques. Comprehensive plans for cleaning up Vincennes and Clichy highlighted the wish to purify suburbs, now regarded as a threat.

The first years of the July Monarchy were a turning point. The cholera morbus epidemic in 1832 made a countrywide disinfection campaign necessary. This campaign marked the beginning of the deodorization of the private space of ordinary people; it acted as a stimulus to sanitary regulation, which had been dormant for some time.

New anxieties were already surfacing. Louis Chevalier rightly emphasizes the growth of new visual demands on the urban milieu during Louis-Philippe's reign.[13] A new collective sensibility took shape among ordinary people. Tuberculosis and the difficulties of gaining access to fresh air strengthened their attitudes. This heightened awareness coincided with a sharp increase in the use of coal, the proliferation of Wilkinson foundries, and the adoption of gas lighting. Thereafter complaints multiplied in Paris against the use of coal (1839),[14] the operation of steam engines, and the opening of bitumen[15] and rubber factories (1836). Smoke aroused concern not so much because of its odor as because it was dark and dense and attacked the lungs, blackened façades, darkened the atmosphere; there was a corresponding demand for more light in the urban environment.

Nevertheless, administrators and experts (whose optimism was now waning) were not impotent. Both experts and police had long thought that high brick chimneys and surrounding walls were enough to render irksome smoke and vapors innocuous. Moreover, new experiments showed that smokeless furnaces were efficient.[16] These appliances put an end to the clouds of smoke emitted by coal burning, tobacco drying, and sugar refining. Not until 1854, however, did administrators make a forceful (though not wholly successful) effort to check the damage from these dark swathes.

The new form of anxiety from the middle of the century onward was reflected in a decrease in the number of references to smell in descriptions of public space. In 1846 Monfalcon and Polinière compiled a detailed list of the "disadvantages" of 213 categories of insalubrious, dangerous, or unpleasant establishments.[17] Quantitative analysis of these disadvantages still showed the primacy accorded in this period to olfactory hazards (cited in connection with 69.4 percent of the categories of establishments);[18] it confirmed continued indifference to noise (2.7 percent) and dust (2.7 percent) but increasing concern about smoke (21.5 percent).[19] Comparison of this document with the list of disadvantages mentioned at the time of the 1866 sanitation decree[20] reveals a slow evolution, with growing attention to noise, dust, and especially smoke.

Second Empire policy translated into action this evolution in sensibilities. Baron Haussmann, who became prefect of the Seine in 1853, set to work to make Paris less dark. His town planning was partly aimed at eliminating the darkness at the center of the city. Although Paris remained foul-smelling—and we have seen why—concern with smells played a declining role in the administration of public space.

If Haussmann's policy could be interpreted—and not unreasonably—as a "social dichotomy of purification";[21] if it was true that a social division of stench—almost uniformly distributed in the not so distant past—was now in force in the city, it was because a switch had taken place over a period of some twenty years. During that period attention to smells and the stench of the poor, along with the indifference of the poor to their own smells, tended to become the essential relay of anxiety about a generalized putrid space. For so long as education had not leveled the thresholds of sensory tolerance,[22] it was thought that purification requirements must be selective, quite apart from the fact that disinfecting the space reserved for bourgeois activities could only enhance property values. Wealth increased when the volume of refuse and the strength of its stench decreased. On the other hand, purifying rented premises crowded with apathetic workers would do nothing for the time being but add inordinately to landlords' expenses. The quest for profit strengthened this social distribution of odors.

PART THREE

Smells, Symbols, and Social Representations

WHILE CHEMISTRY, as a result of Lavoisier's discoveries, revolutionized perceptions of space and caused the old theories of air to be discarded, a theoretical movement emerged to give olfactory sensations new significance and to confer unexpected prestige on olfactory messages despite the shortcomings of osphresiological knowledge.

In the nineteenth century Cabanis leveled a powerful, if partial, criticism against this sensualist approach derived from Locke. "Good analysis," he wrote, "cannot isolate the operation of any one particular sense from the operations of all the others." The "senses [are] in continual reciprocal dependence." The odor of the rose acquired some of its characteristics "through the conjunction of other simultaneous sensations." Above all, "as each sense is able to come into operation only by virtue of the earlier operation of all the general systems of organs, and able to continue only by virtue of their simultaneous operation, it always necessarily feels the effects of their behavior and shares to a greater or lesser degree their most ordinary defects. Thus the degree of sensibility of the sensory system and its counterbalancing relationships with the motor system have considerable influence on the character of the impressions received by each particular sense."[1]

The sense of smell, for its part, maintained "intimate relation-ships" with several organs; indeed, it became accepted as the sense of affinities. Its close liaison with the sense of taste had already been acknowledged; a liaison was also established, by Cabanis, between the nose and the intestinal canal; and several stomach ailments caused anosmia. A century before Fliess,[2] Cabanis also emphasized a phe-nomenon that later provoked lengthy polemics: the relationship be-tween the olfactory membrane and the genital organs.

Cabanis sought a new treatise on sensations, based on the rela-tionships among the sensory mechanisms and among the sensory mechanisms and the other organs. This "physiological history of sen-sations,"[3] in practical terms not so far removed from the "science of the sensitive being" to which Maine de Biran devoted himself in his *Journal,* opened up new perspectives. As a result osphresiology, even as it was developing, lost its relevance and lay dormant for a long time.[4] The Restoration doctors' untiring refutation of Condillac's rigid epistemology,[5] together with the belated revival of vitalism, hastened this disaffection.

But Cabanis, who considered that "individual life is in sensa-tions," at the same time hallowed the sense of smell as the sense of sympathy and antipathy among beings.[6] Like the osphresiologists, he emphasized the specificity of individual odors and atmospheres. It was no longer enough to stress variations in body odor according to age, sex, or climate; the person's atmosphere and his behavior toward smells revealed his individuality. There was no "organ with more individual sensations" than the nose, asserted Dr. Fournier in the *Dictionnaire des sciences médicales.*[7]

The refinement of this sense varied according to habitat. "People who live within the bonds of society," Virey noted, "are more af-fected by vegetable odors, whereas the savage smells the putrid fe-tidities of animal bodies much better."[8] Civilization, Dr. Kirwan added later, made strong odors intolerable and dangerous.[9] Anthropology, like medicine, justified the fashion for vegetable scents and the de-cline in animal perfumes.

Anthropology also helped explain another paradox, the fact that sensitivity to delicate, pleasant odor, which implied a specific access to smells and therefore the deodorization of the environment in general, developed in an inverse direction to the ability to analyze odors, which presupposed a lengthy apprenticeship. "The inhabitants of Kamchatka," noted Virey, "could hardly smell spirituous melissa cordial or eau de cologne at all, whereas they could catch the scent of rotten fish or a grounded whale very well from a distance."[10]

Similarly, the heavy laborer, plunged day and night in a vitiated atmosphere, impregnated with heavy odors, entirely taken up with manual work, his person exhaling heavy scents, lost his access to the sense of smell. By the law of compensation governing the development of organs, strong arms precluded a delicate nose; the latter remained the prerogative of people who were not forced to perform manual labour. Unequal sensibilities were only one more sign of inequality among individuals.[11]

All these beliefs laid the foundations for what I shall quite improperly call the bourgeois control of the sense of smell and the construction of a schema of perception based on the preeminence of sweetness.

The delicacy of an individual's atmosphere and the sensitiveness of his sense of smell were evidence of his refinement and proved his ignorance of the sweat of hard labor. This acuity could even become excessive and dangerous; delicate young girls, for example, might fall victim to parosmia (confusion of smells). Olfactory messages assumed great importance in this protective and preternaturally aware world. The sense of smell administered pleasures, the innocence of which was guaranteed by refinement.

Once again, the history of perception reveals its contradictions. Whereas chemical analysis tended to replace investigation based on the senses, and while osphresiological research lagged behind, olfaction was caught up in the refinement of nineteenth-century practices and divisions. The subtle interplay of individual, familial, and social atmospheres helped to order relationships, governed repulsions and affinities, sanctioned seduction, arranged lovers' pleasures, and at the same time facilitated the new demarcation of social space.

9 The Stench
of the Poor

The Secretions
of Poverty THE INCREASED attention to social odors was the major event in the history of olfation in the nineteenth century before Pasteur's theories triumphed. Whereas references to stenches from the earth, stagnant water, corpses, and, later, carcasses gradually diminished, the discourse of public health and the language of novels, as well as nascent social research, spoke of smells to the point of suggesting an obsession with a human swamp. The shift from the biological to the social was a result of Cabanis's project. Moreover, observers no longer analyzed only the smells of hospitals, prisons, and all those sites where people confusedly crowded together to produce the undifferentiated odors of the putrid throng. A new curiosity impelled them to track down the odors of poverty in the very dens of the poor.

This reorientation away from public toward private space required a complete change of strategy. "While continuing to stress the utility of broad streets, houses with good aspects, cleanliness of villages, drainage of swampy lands, [we] state that it is not the outside wall, but the actual inhabited room itself where the greatest watch on salubriousness has to be kept," Piorry concluded after reading the reports on epidemics in France between 1830 and 1836.[1] Passot summarized this view brilliantly fifteen years later: "The wholesomeness of a large town is the sum of all its private habitations."[2]

The new effort to monitor stench inside the dwellings of the humble was inseparable from the development, among the bourgeoisie, of a system of perceptions and a model for behavior in which olfaction was only one component, though not a minor one. The sudden awareness of the growing differentiation of society was an incentive to refine analysis of smells.[3] Other people's odor became a decisive criterion.[4] Charles-Léonard Pfeiffer, for example, has shown what skillful detail Balzac used in *La Comédie humaine* to locate the status of bourgeois and petit-bourgeois, peasants or courtesans, by the odors they emitted.[5]

Once all the smells of excreta had been got rid of, the personal odors of perspiration, which revealed the inner identity of the "I," came to the fore. Repulsed by the heavy scents of the masses— symptomatic of how hard it was for ideas of differentiated individuality to emerge in that milieu—impelled by the prohibitions on the sense of touch, the bourgeois showed that he was increasingly sensitive to olfactory contact with the disturbing messages of intimate life.

The social significance of this behavior is flagrantly obvious. The absence of intrusive odor enabled the individual to distinguish himself from the putrid masses, stinking like death, like sin, and at the same time implicitly to justify the treatment meted out to them. Emphasizing the fetidity of the laboring classes, and thus the danger of infection from their mere presence, helped the bourgeois to sustain his self-indulgent, self-induced terror, which dammed up the expression of remorse. From these considerations emerged the tactics of public health policy, which symbolically assimilated disinfection and submission. "The enormous fetidity of social catastrophes," whether riots or epidemics, gave rise to the notion that making the proletariat odorless would promote discipline and work among them.[6]

Medical discourse went hand in hand with this evolution in sensory behavior. Shaken by anthropology and the nascent empirical sociology, medical science let some fundamental neo-Hippocratic principles fall by the way. Topography, the nature of the soil, climate, and the direction of winds gradually ceased to be regarded as determining factors; experts emphasized more than ever the harm caused by crowding or proximity to excrement; above all, they now accorded decisive importance to the "secretions of poverty."[7] This was basically the conclusion of the report on the 1832 cholera morbus epidemic.[8]

Doctors and sociologists had just detected that a type of population existed which contributed to epidemic: the type that wallowed in its fetid mire.

It is now easier to understand the persistent anxiety aroused by excrement. The ruling classes were obsessed with excretion. Fecal matter was an irrefutable product of physiology that the bourgeois strove to deny.[9] Its implacable recurrence haunted the imagination; it gainsaid attempts at decorporalization; it provided a link with organic life, as the traces of its immediate past. "We find the candor of refuse pleasing and restful to the soul," confessed Victor Hugo, alive to the history that could be read in waste.[10] Parent-Duchâtelet and many others set out to explore the mechanisms of the necessary evil of urban excretion from an organicist and Augustinian viewpoint. Crossing the center of the city, they met the men who worked with filth. Excrement now determined social perceptions. The bourgeois projected onto the poor what he was trying to repress in himself. His image of the masses was constructed in terms of filth. The fetid animal, crouched in dung in its den, formed the stereotype. The bourgeois emphasis on the stench of the poor and the bourgeois desire for deodorization were therefore inseparable.

This new attitude was a departure from eighteenth-century anthropologists' fascination with the odor of bodies, which did not connect it with the state of poverty but attempted to relate it to climate, diet, profession, and temperament. These pioneers analyzed the odor of the old man, the drunkard and the gangrenous, the Samoyed and the stableboy, but rarely the poverty-stricken. The fetidity of the throng was dangerous only because of the crowding and mingling of people. At the most, Howard declared that the air surrounding the poor man's body was more contagious than the air surrounding the rich man's, but he made no reference to a specific stench.[11] He only implied that disinfection techniques had to be modified according to degree of wealth.[12]

Nevertheless, medical science in those days suggested that some individuals exhaled an animal stench. The human who had always wallowed in the depths of poverty smelled strong because his humors did not have the necessary digestion and the "degree of animalization proper to man."[13] Therefore, if he did not have a human odor, it was not because he had regressed but because he had not crossed the threshold of vitality that defined the species. Accordingly, portraits of madmen and some convicts reflected the model of a chained dog squatting in a trough, turning its bed into a dunghill, and dripping urine like a liquid manure sump. These portrayals gave birth to the image of the "dung-man," impregnated with excrement, forerunner of the image of the foul-smelling laboring proletariat of the July Monarchy.[14]

As early as the eighteenth century, several other groups had a similar image. Foremost among these, it goes without saying, were prostitutes, typically associated with filth and whose presence diminished when refuse disappeared from the streets. In Florence, Chauvet noted that the streets were paved, drains covered, rubbish contained behind screens, "roads strewn with odoriferous flowers and leaves";[15] there was no longer a single prostitute to be seen.

Jews also were regarded as filthy individuals. They owed their unpleasant odor, it was said, to their characteristic dirtiness. "Everywhere these Hebrews gather," Chauvet asserted, "and where they are left to administer their precinct themselves, the stench is singularly perceptible."[16]

The ragpicker brought the linkage between unpleasant odor and occupation to its extreme, because his person concentrated the malodorous effluvia of excrement and corpses.[17] Domestic servants also smelled unpleasant, although their status and hygiene improved. In 1755 Malouin advised airing as much as possible the places where they had been.[18] In 1797 Hufeland ordered their exclusion from children's nurseries.[19]

Between 1800 and the aftermath of the great cholera morbus epidemic in 1832, the image of Job, in the guise of the dung-man, became linked to the obsession with excrement. A favorite subject of early, flattering social research was the city's untouchables, the comrades in stench, the people who worked with slime, rubbish, excrement, and sex: sewermen, gut dressers, knackers, drain cleaners, workers in refuse dumps, and dredging gangs attracted the attention of the early pioneers of empirical sociology. I have emphasized elsewhere the immense epistemological significance of the inquiry into public prostitution in the city of Paris, which claimed Parent-Duchâtelet's attention for nearly eight years.[20] The archives of the conseils de salubrité confirm this special interest.

In addition to the evidence on prostitutes, there are other examples. The convict wallowing in his filth was still an inexhaustible theme.[21] Certainly, in the eyes of contemporary theorists, this figure of the convict had become virtually irrelevant. Nonetheless, studies of the penitentiary bear witness to a continuing reality. Dr. Cottu described his visit to a dungeon in Reims prison: "I felt I was being stifled by the horrible stench that hit me as soon as I entered . . . At the sound of my voice, which I tried to make soft and consoling, I saw a woman's head emerge from the dung; as it was barely raised, it presented the image of a severed head thrown onto the dung; all the rest of this wretched woman's body was sunk in excrement . . . Lack of clothing had forced her to shelter from the stringencies of the weather in her dung."[22]

In 1822 alone the ragpicker, the archetype of stench, was the subject of seventeen reports by the Conseil de Salubrité.[23] The authorities tried to move away from the city the malodorous dumps where, prior to sorting, he piled up the bones, carcasses, and all the remains collected from public highways. The council looked kindly only on collectors of "bourgeois rags"; there was no danger that these would transmit the infection of the masses. The ragpicker concentrated the odors of poverty and was impregnated with them; his stench acquired a symbolic value. Unlike Job or the putrefying convict, he did not wallow in his own dejecta; the grimacing face of the rubbish of the masses, he sat on other people's dung.

Down the rue Neuve-St.-Médard, rue Triperet, or, even more, the rue des Boulangers, individuals were to be seen "dressed in rags, without shirts, stockings, or often shoes, crossing the streets whatever the weather, often going home soaked . . . laden with different products plucked from the capital's refuse, its fetid odor seeming to be so much identified with their persons that they themselves resemble veritable walking dunghills. Can it be otherwise in view of the nature of their activity in the streets, their noses continually in dunghills?"[24] When they got home they sprawled on stinking and dirty straw amid vile-smelling refuse.

Blandine Barret-Kriegel discerned an element of fascination in the shocked gaze of those who visited the poor—from Condorcet to Engels, from Villermé to Victor Hugo—for "the ragpicker's dustbin house," "infernal dwelling," for "the unpleasant smell of another, more barbarous, stronger life," the "eternal return of subterranean powers."[25] Accounts of behavior toward smells along with frequent references to the stenches of hell confirm that, whether it concerned excrement, prostitutes, or ragpickers, the fascination mixed with repulsion pervaded the discourse and governed the attitude of sanitary reformers and social researchers.

It almost goes without saying that homosexuals shared in the stench arising from intimacy with filth. Symbols of anality, congregating in the vicinity of latrines, they also partook of animal fetidity.[26] According to Félix Carlier, the odors of the pederast, who was addicted to heavy perfumes, showed the close relationship between the smells of musk and of excrement.[27]

The case of the sailor has been less closely investigated. As the ship, stockpot of every stench, quickly became the laboratory for experiments with ventilation and disinfection techniques, the individual who lived on board was to be seen as a necessarily important object of inquiry. He, after all, ran the greatest risk of falling victim to vile-smelling effluvia, as the tragic fate of the *Arthur* showed. Authors of manuals on maritime health were categorical: the sailor smelled unpleasant and was disgusting. "His customs are debauched; he finds supreme happiness in drunkenness; the odor of tobacco, wedded to the vapors of wine, alcohol, garlic, and the other coarse foods that he likes to eat, the perfume of his clothing often impregnated with sweat, filth, and tar make it repulsive to be near him." The stench of the sailor, "robust and libidinous," condemned to long continence or masturbation, added a strong spermatic secretion to the effluvia.[28]

Fortunately, sailors—and in this context the crew stood for the masses at large—did not have a good sense of smell. They did not share the officers' revulsion, because they lacked delicate noses. Dr. Itard had stated that Aveyron's savage child felt no disgust for his own excrement.[29] The link that sanitary reformers established between stench and the relative anosmia of the masses only confirmed the bourgeoisie in their push toward deodorization. Although sailors were admitted to be keen-eyed, "hearing presents a slight difficulty" because of the uproar from storms and artillery; "the sense of smell is insensitive in that it is little exercised; the roughness of manual work makes the sense of touch very dull; the sense of taste is depraved by gluttonous and unrefined appetites."[30]

In general "the sailor's sensory organs enjoy little activity; the nerve ends seem to be hardened by rough physical work and paralyzed by lack of exercise of the intellectual faculties."[31] He would probably be unresponsive to the balsamic odors of spring flowers; far from the sights of rural nature, "his senses are no longer fine enough to analyze its charms."[32] Worn out by strong emotions, sailors were unable to experience refined feelings. The sensory inferiority, not to say disability, of the masses engendered a corresponding poverty of ideas and of feelings. Conversely, the refinement and acute sensitivity of the officer threw the deterioration of the sailor into even sharper relief, and justified the respect shown by the crew. After the cholera morbus epidemic, when the moral calculus movement was renewed, proletarian poverty became the favorite subject of social research. Now denunciations were directed against the stench of the masses as a whole instead of against a few isolated categories symbolically identified with excrement. If servants, nurses, and porters smelled unpleasant, it was because they brought the odor of the proletariat into the bosom of the bourgeois family; this was enough to justify their exclusion, with the exception of the wet nurse.[33] The neurotic Flaubert was a privileged witness to this repulsion toward "the basement odor" that emanated from the masses. "The journey back was excellent," he wrote to Madame Bonenfant on May 2, 1842, "apart from the stench exhaled by my neighbors on the top deck, the proletarians you saw when I was leaving. I have scarcely slept at night and I have lost my cap."[34] Huysmans later carried this intolerance of smells to its logical conclusion.

Jacques Léonard's linguistic analyses of medical discourse emphasize how frequently the terms *wretched, dirty, slovenly, stench,* and *infect* were used together.[35] The unpleasant odor of the proletariat remained a stereotype for at least a quarter of a century, until the attempts at moralization, familialization, instruction, and integration of the masses began to bear fruit. Air, light, a clear horizon, the sanctuary of the garden were for the rich; dark, enclosed areas, low ceilings, heavy atmosphere, the stagnation of stenches were for the poor. The archives of the conseils de salubrité and the Constituent Assembly's 1848 inquiry into agricultural and industrial work are the crucial texts in this endlessly recycled discourse.

Several images dominated descriptions of poverty. Like the stench of certain artisans not long before, the stench of the poor man was attributed to impregnation even more than to his carelessness in disposing of all his excreta. Like earth, wood, and walls, the worker's skin and, even more, his clothing, soaked up foul-smelling juices. In the Pompairin spinning mill, Dr. Hyacinthe Ledain wrote, the children were rickety. "Their condition is attributed to the fact that the air they breathe is unhealthy as a result of the large quantity of greasy oil used in these establishments. The clothing covering these children is so impregnated with it that the strongest, most repulsive odor can be smelled when they approach." The Secondigny textile mill was just as unhealthy. The children were hideous. "They can be seen coming out of their workshops covered in rags impregnated with oil."[36] Jacques Vingtras felt repulsion for the lamplighter at the college at Le Puy who exhaled an odor of machine oil.[37] Again, in 1884 Dr. Arnould declared that the Lille poor were "inferior to the rich, not because of work, but because of their narrow, sordid shelters [the poor did not have dwellings], the uncleanliness that surrounds and penetrates them, their life in contact with filth which they have neither the time nor the means to get rid of and which even their education has not taught them to fear."[38] While carrying out his retrospective research into working conditions in the north of France on the eve of World War I, Thierry Leleu heard it said that the reel-girls, called "chirots" (a dialect form of *sirops,* "syrups") because of the liquid that poured out of the machines, "had the odor of linseed gum. A girl who worked in a spinning mill could be recognized by her odor, even in the street. This odor stuck to her skin."[39] Popular novels later conveyed this perception and the repulsion it aroused. Their descriptions of factories stress the stench and suffocating heat rather than the industrial processes.[40]

The odor of rancid tobacco that impregnated workingmen's clothing was another common theme.[41] Everything seems to suggest that the effluvia from tobacco was tolerated to only a limited extent at the end of the eighteenth century—the ruling classes were probably more tolerant of farting and the odor from latrines. Tobacco—pipe, cigar, then cigarette—conquered public places in the first half of the nineteenth century. At first glance this phenomenon seems to run counter to the strategy of deodorization then in progress; however, some doctors still attributed disinfectant properties to smoke. Old soldiers, veterans, and halfpay officers, as well as sailors, were responsible for its spread.[42]

From this point on, tobacco never lost its ambiguity. Its odor signaled the arrival of the boor;[43] the majority of sanitary reformers denounced it. Michelet accused it of killing sexual desire and reducing women to solitude; Adolphe Blanqui demanded that women and children be forbidden to use this drug, because "it is the beginning of every disorder."[44]

The repulsion often assumed a sociological significance. Forget reviled the sailor's quid; its odor impregnated his breath, hands, and clothing. It was true, he observed on a conciliatory note, that it was a form of compensation; it should therefore be tolerated. "The sailor uses tobacco as you use coffee, balls, and entertainments, as the literary man feasts on Voltaire, the scholar on an abstract problem."[45] "Tobacco is the only thing that assists the imagination of the poor," pleaded Théodore Burette in his *Physiologie du fumeur*.[46]

But tobacco's victory also symbolized the victory of liberalism; it bore witness to increasing male domination of social life before it actually became its instrument. Like conscription, to which its spread was largely due, tobacco was decked with "patriotic," egalitarian qualities. It was in this context that it earned its title to nobility. "Smoking creates an equality among its confraternity . . . rich and poor rub shoulders, without being surprised by the fact, in places where tobacco is sold," and only there.[47] "The firmest support of constitutional government,"[48] the July Monarchy ensured its triumph. For our purposes, it is important to note that this successful popularization occurred at exactly the time that the stench of the laboring classes was perceived as a natural act of the social landscape.

The stress on the repulsive smell of the proletariat appears clearly in the accounts by doctors and visitors to the poor. This was a new intolerance. Hitherto, doctors had seemed impervious to disgust; only fear of infection appeared to motivate precautions.[49] During the second third of the century, repulsion toward the smell of the masses was openly acknowledged, without any real recognition whether this represented a new intolerance or a new frankness. The patient's domicile became a place of daily torture for the doctor. "One positively suffocates there," Monfalcon and Polinière stated. "It is impossible to go into this center of infection; often the doctor who visits the poor cannot bear the fetid odor of the room; he writes his prescription by the door or the window."[50]

Unlike his wretched clientele, the doctor no longer tolerated animal effluvia. "On entering this house," noted Dr. Joire in 1851, "I was struck by the foul-smelling odor breathed there. This odor was literally stifling and unbearable and seemed like the smell of the most fetid dung; it was particularly strong around the patient's bed, and was also spread through the whole apartment, despite the outside air that came in through the half-open door. I could not remove from my nose and mouth the handkerchief with which I protected myself the whole time I stayed with this woman. Yet neither the inhabitants of the house nor the invalid seemed to notice the inconvenience of the miasma."[51] Adolphe Blanqui, assailed by the stench of the Lille cellars and by the odor of filthy men emanating from them, recoiled in shock at the entrances to these "ditches for men"; only in the company of a doctor or a police officer did he "hazard" descent into this hell where "human shadows" tossed and swirled.[52]

Inside the workshop, on the ship's bridge, in the sickroom, the threshold of perception, or more precisely of tolerance of smells, defined social status. Bourgeois repulsion accompanied and justified phobia about tactile contact. The patient's stench rather than respect for feminine modesty established the use of the stethoscope.[53]

This social division perceived through bodily messages also embraced the disgust inspired by teachers, schoolmasters, even professors. Paul Gerbod skillfully demonstrated that their image then was that of an antihero.[54] These old, frustrated bachelors, whose former bourgeois pupils remembered their odor of sperm and rancid tobacco, had proved unable to fulfill their dreams of promotion; their stench, like the stench emitted by clergy of humble descent,[55] continued to betray their origins.

The masses gradually came to feel the same repulsion. The new sensitivity reached that fringe of workers who spent their nights trying to escape being haunted by their involvement in manual labor. Hitherto unperceived horrors had to be endured in the process of adopting the new culture. The warm consolation of sleeping more than one to a bed had to be given up. Norbert Truquin, a railway navvy, felt his gorge rise when he breathed the odor of brandy and tobacco exhaled by his companions; forced to share his pallet, he confessed that he could no longer without repulsion tolerate contact with another man.[56]

Cage and Den

The flood of discourse on the habitat of the masses and its stifling atmosphere revealed the new preoccupation after the 1832 cholera morbus epidemic. "The atmospheric swamp of the house" had replaced the cesspools of public space in the hierarchy of anxieties about smells.[57] In towns, complaints concentrated on the stench of the communal sections of the dwellings of the masses. The basic theme of the diatribe was a denunciation of the odor of excrement and refuse, which had not yet been privatized in these sections of society. Consequently the denunciation of stench was closely linked to the denunciation of promiscuity. On this subject, discourse on sanitary reform was churned out with tedious monotony. Lachaise, Hatin, Bayard, Blanqui, Passot, Lecadre, Tetrais, Ledain, and many others unflaggingly copied each other or just repeated themselves. It would be interesting to analyze in detail the operation of this obsessional litany from the point of view of psychohistory. Popular novels, as Marie-Hélène Zylberberg-Hocquard has shown, used these shocked descriptions of vile-smelling homes for their own ends; this is not surprising, since the novelists were inspired by the writings of social researchers.[58]

The odor of stagnant urine, congealed in the gutter, dried on the paving, encrusted on the wall, assailed the visitor who had to enter the wretched premises of the poor. The only means of entry was through low, narrow, dark alleys. These formed channels for a fetid stream laden with greasy water and "the rubbish of every type that rained down from all the stories."[59] Gaining access to the poor man's stinking dwelling almost amounted to an underground expedition. Adolphe Blanqui moved through the Lille courtyards or the Rouen slums with the fascinated caution that earlier had driven Parent-Duchâtelet to cross the sewers of the city. The narrowness, darkness, and humidity of the small inner courtyard into which the alley opened made it look like a well, the ground carpeted with refuse. Here, garbage rotted, laundry- and dishwater formed pools; stenches amalgamated and rose to nourish the fetidity of the upper stories. Within this order of perception the staircase acted as an overflow; a foul-smelling cascade rushed down it, checked at every floor by the landing, fed by the latrines, which revealed through open doors the obscenity of the privy full of excrement. Dr. Bayard retained the aural memory of the "gurgling household water" on staircases in the IVe arrondissement of Paris.[60] The stench of these premises formed an ensemble. The odor of excrement predominated; it varied only in strength from one place to another. There was no subtle division into different categories of smells here.

Inside the dwelling, congestion, a jumble of tools, dirty linen, and crockery, prevailed. The poor man "wallowed" amid this disorder, often in the company of animals;[61] the cage rather than the den was the dominant image. "Poverty is enclosed in a narrow dungeon."[62] The obsession with air now focused on the dungeon; its lack of air seemed all the more apparent since scientists had succeeded in defining precise norms of ventilation. More than conveying the presence of miasma, stench now threatened suffocation. This basic psychological shift helps explain the forms the new vigilance took.

In particular, writings focused on the aspect of narrowness. The crampedness of the sleeping area, the depth of the yard, and the length of the alley created in the mind of the bourgeois (who normally had plenty of room) the impression of suffocation. The phobia about lack of air focused attention on the stifling atmosphere of the craftsman's garret under the roof, the low ceiling of the lodge where the porter crouched like a dog, the tradesman's back shop, and the narrow closet of the student or draper's assistant.

Lodgings and hostels were even worse. Louis Chevalier has noted the repulsion aroused by the smell of immigrants from the provinces.[63] City dwellers' repugnance and contempt for the regional odors that impregnated seasonal workers from Limoges or Auvergne helped to justify the segregation of these country people for a long time.[64] Martin Nadaud was retrospectively shocked at the indifference that masons from the Creuse showed to the stench of their hostels. Both the vicomte d'Haussonville and Pierre Mazerolle denounced the odors from the cheese and bacon piled up on the shelves there.[65]

The dormitories for the people of the Limousin region were well organized; yet the confusion that reigned in certain overnight lodgings haunted the bourgeois imagination. Visitors were aghast when faced with this total promiscuity. Here, the brotherhood of filth fostered an atmosphere of animality. Individuals, it was said, coupled freely there.[66] "Did these people really know each other?" asked Victor Hugo about the *Jacressade*'s imaginary guests. "No, they sniffed one another."[67]

"Rooms that are too narrow for one man to live in produce effects just as deadly as spacious chambers where many men are gathered," wrote Piorry about the dwellings of the masses.[68] In this environment, the sickroom recreated the marsh. It combined all the conditions of the swamp in the equatorial jungle, stated Dr. Smith.[69] This was where those putrid fevers incubated that it was eventually suggested might be the result of slow asphyxia, accompanied by ataxia and adynamia.[70] The unpleasant odor was evidence of the lack of air that hampered the efficient deployment of the work force. What was described as shameful laziness was most often only "debilitation . . . from the vitiated atmosphere in unhealthy dwellings."[71] The poor had to be given air; doctors and sanitary reformers were unanimous on this point. Ventilation and deodorization were economic imperatives. Gabriel Andral, Louis, Jean Bouillaud, Chomel, and many others carried out abundant observations to measure the effects of congestion, which, according to Jean Louis Baudelocque, was the cause of scurvy. Studies of cholera had established "the almost constant relationship between the gravity of the symptoms and the tininess of dwellings"; it was probably the smallness of the dwelling that gave the disease its "typhohemic and mortal character."[72] Villermé focused on the ravages wrought by cholera in lodgings; the most congested areas were the most deadly.

The sense of smell was still better than the instruments of physics for measuring the renewal of air and thus for averting the ill effects of overcrowding. But there was increased concern for light in private dwellings, as in public space; this was the beginning of the great swing in attitudes that was to give uncontested supremacy to the visual. Moreover, Baudelocque noted that dark places made flesh soft, puffy, and flaccid; inadequate light slowed circulation, brought on the young girl's terrible chlorosis; Jean Starobinski has stressed its effect on the imagination.[73] Darkness made nocturnal animals sad and perfidious; uncertain light was a threat to health, zeal for work, and sexual morality.[74] A young husband's first duty, stated Michelet, was to give his child and its young mother "the joy of a good light."[75]

The countryside had never been thought particularly sweet-smelling. The peasant's lack of hygiene and the strong odor of his sweat were very old themes; Sancho Panza had daydreamed about the heavy scent of Dulcinea's armpits,[76] and a couple of centuries later Rousseau's contemporaries complained freely on the same theme. The agitation about excrement had not been confined to the towns; it penetrated to the depths of the countryside. Country smells were often the objects of attack. In 1713 Ramazzini had already denounced the foul-smelling proximity of dung and, even more, the horrible stench from steeping hemp.[77] Before the discoveries by Priestley and particularly Ingenhousz, people were afraid of being near trees, lest these add to the ill effects of the subterranean blasts that beset the laborer. Even the air from kitchen gardens, stinking of manure, concealed many dangers. Like swamps, villages engendered miasma.[78]

All this is a far cry from Julie's garden and Jean-Jacques's reveries. Two apparently contradictory systems of perception were intermingled; as a result of this duality, the image of the countryside remained complex throughout the following century.[79] For the moment, the contradiction was only on the surface. The countryside exalted by

Rousseau and his disciples emerged as a sweet-scented area, free of stenches from the village and its assembled peasants, wafted by nothing but the breath of spring flowers. It was a countryside that seemed to have been created for solitude, where the traveler seemed able to tolerate only the isolated farm, the mill, the chalet, at a pinch the hamlet, and the momentary contact of a chance meeting with a shepherd.

This idyllic vision of the peasant and the life of the fields survived into the nineteenth century. Picturesque journeys, and particularly iconography, helped to keep it alive.[80] Unlike the everyday contact of medical practice, which involved the senses of touch and smell, ethnology via observation allowed distance; it permitted scales of revulsion. The artist's brush easily transferred reality into symbolism.

Nevertheless, the village was soon perceived as the antithesis of the mountain summit bathed in the purity of the ether, and was painted in dark colors. Social emanations fermented in the depths of valleys; travelers should not leave the slopes of hillsides. Obermann fled from low ground; Dr. Benassis set out to curette it. This was no hopeless undertaking: as early as 1756, Howard successfully transformed Cardington peasants' "mud huts," where in his opinion they lived like savages, into cheerful cottages.[81]

Charles-Léonard Pfeiffer has cataloged the manifestations of Balzac's repugnance to the smell of peasants. Here is just one example: "The strong, savage odor of the two habitués of the highway made the dining room stink so much that it offended Madame de Montcornet's delicate senses and she would have been forced to leave if Mouche and Fourchon had stayed any longer."[82]

When Balzac wrote *Le Médecin de Campagne* (1833) and *Les Paysans* (1844) the stench from villages had been feeding a steady stream of writing for some years. No report read to the Conseil de Salubrité from whatever rural department, no medical thesis about the peasant environment, no report on an inquiry under the July Monarchy or the Second Republic failed to denounce violently the poor hygiene of the habitat of rural space. Thus, every book about the social history of the French countryside at that time gives considerable space to this complaint.[83] Most of the authors—including myself—have rather naïvely used the copious discussions by bourgeois observers for their own purposes. It would have been more valuable if they had tried to unravel the tangled systems of images and, above all, shown that the basic historical fact was not the actuality (which had probably changed little) but the new form of perception, the new intolerance of traditional actuality. This sensory change within the elite and the flood of discourse it provoked were to bring about the revolution in public health, the road to modernity.

A reversal thus took place at the level of perceptions. Filth and rubbish, so greatly feared by refined city dwellers, invaded the image of the countryside; even more than in the past, the peasant tended to be identified with the dung-man, intimate with liquid manure and dung, impregnated with the odor of the stable. Hitherto, the public stenches of the town had been under fire; now, the town was—slowly—cleared of its refuse; half a century later it had almost succeeded in cleaning up its poor. Its relationship with rural space was reversed: it became the place of the imputrescible—that is, of money—whereas the countryside symbolized poverty and putrid excrement.[84] The power of agrarian ideology was not sufficient to challenge a perceived reality, which the negative welcome given to immigrants from the countryside and the attitude of travelers or city-dwelling tourists bore out for more than a century.[85] A new relationship between the images of town and countryside was not established until the arrival in the latter of water supply, mechanization, household equipment, and ecological propaganda.

In their repetitive descriptions, the explorers of peasants' households under the July Monarchy confined themselves to a few stereotypes that make the discourse wearisome to the reader. This monotony had set in as early as 1836, as Piorry's analysis shows.[86] The cramped nature of the premises, the narrowness of the windows, the lack of air and light, the dampness of the floor aggravated by the absence of paving, the ill effects from smoke, the stench of dung coupled with the odors of laundry and washing-up, the exhalations of putrid and fermented scents from stable and dairy located too close by were the basic elements of the picture. The use of deep featherbeds that became impregnated with the sleeper's sweat, the presence of domestic animals, competing with men for air, and the numerous hams hung from the ceiling added substance to witnesses' complaints; only rarely did they deplore inadequate bodily hygiene. They were obsessed with the animal stench of the place, not yet with its lack of refinement. The normative system being built up elsewhere could not yet be applied to the peasant;[87] all he was asked to do was to remove his dung and poultry droppings, then to open his door and windows wide.

The stench of the poor became less of an obsession during the second half of the century. Advances in hygiene caused the concern to become more specialized and marginal in its choice of object. Peasants became the objects of a repulsion which was already obsolete and which veered toward barrack-room humor.[88] The same was true of seasonal workers, maids, porters, and a few workers in particularly dirty urban trades (such as the "chirots" in the north). The description of the backstairs in Zola's *Pot-Bouille* exemplifies this obsession; it shows the unwelcome presence of people who could hardly be regarded as a serious threat any longer.

Tramps and vagabonds were endowed with a specific stench, and this development proved that the proletariat had lost some of its threatening smells. According to the Goncourt brothers, the odor of cockchafers was "recognized at the prefecture as the special odor of the vagabond, the man who sleeps under bridges; the odor of the convict and the prisoner."[89] We are thus returned to the odor of the dungeon; the circular form of perception in which the confused stench of the proletariat becomes complete. Henceforth it was the odor of race that would constitute a threat and provide the focus of scholars' attention.[90] But that is another story.

Cleaning Up the Wretched

Under the July Monarchy, the perception of the stench of the poor required their deodorization or, alternatively, their disinfection. The objective was to abolish the vile-smelling organic odor that bore witness to the presence of death and could provoke a return of that "brain fever"[91] so murderous in the recent past. Durkheim was to insist on distinguishing the moral element from society's preoccupation with hygiene,[92] but before him the moral implications of the public health venture were emphasized on many occasions in the seventeenth and eighteenth centuries; they were particularly apparent under the July Monarchy. To rid the masses of their animal fetidity, to keep them at a distance from excrement, was part of the therapeutic strategy deployed against social pathology. When stench declined, violence was blunted. Hygiene reigned supreme "against the vices of the soul . . . a crowd with a liking for cleanliness soon has a liking for order and discipline," wrote Moléon, chairman of the Conseil de Salubrité, as early as 1821.[93] "Cleanliness," de Gérando remarked in 1820, "is simultaneously a means of preservation and a symptom that betokens the spirit of order and preservation; it is distressing to see how unknown it is to most poor people, and this is a sad symptom of the moral disease that afflicts them."[94]

Twenty years later, Monfalcon and Polinière still entertained the fantasy of the odorless worker: "After the respiration of pure air, cleanliness, temperance, and work are the principal conditions for the well-being of the laboring classes"; the dwelling of the good worker "has no luxury, but nothing in it injures the senses of sight or smell . . . Merely because this worker breathes a sufficient quantity of healthy air and has plenty of water for his daily needs, he is in better health and earns more. Content in his domicile, he has more respect for cleanliness and the law and is more devoted to the observance of his duties."[95] The indefatigable artisan did not smell strongly, and Zola, in love with Pauline, extolled "the healthy odor of her housewife's arms."[96]

However, there was no question of bathrooms for the time being, and bodily hygiene was limited to a few very specific occupations. Baths were taken almost solely by miners and furnacemen, soiled by coal dust, and by some domestic servants in close contact with the elite. The aim was to remove grease and dirt and, most of all, to wash the face. Of overriding importance was the battle against impregnation of clothing. Being clean meant, above all, having clothing that was free of grease and odor.[97] Thus for a long time the first injunction of what the masses called "bodily" hygiene was to have their personal belongings cleaned. According to Cadet de Vaux in 1821, the crust of dirt covering her clothes, combined with the coarseness of her chemise, prevented the woman of the people from giving off her personal atmosphere and deprived her of the basic element in her power of attraction.[98]

In the town, overcoming the uncleanness of communal conveniences and draining off the filth from the courtyards seemed the most urgent requirements. Progress occurred via the semiprivatization of latrines and the distribution of keys to families whose dwellings opened onto the landing.[99] In this environment the advance in "privacy" consisted chiefly in protection against other people's dirt and odors, in the achievement of an approximate familialization of excrement, and in protecting modesty against potential dramatic interruption. Abolishing the promiscuity of latrines, keeping doors closed, and installing blowoff pipes were indispensable preliminaries for that disciplined defecation deemed essential to the elimination of stenches. It was also important to keep watch against individuals who urinated in alleys: this was a task for the good porter. If necessary, Passot noted, he could set up a small barrier outside and cover the gutter with a slab.[100] In short, the venture aimed at progressively transforming communal conveniences into private conveniences. Frequently, whitewashing and painting to eliminate impregnation of walls completed the arsenal of measures advocated. Obviously, the advance involved a subscription to the water company; the manifold obstacles that blocked the extension of this practice are well known.

In the country and in a number of small towns, the struggle against the stench of excrement sustained the interminable battle between municipal officials on the one hand and the owners and users of dung on the other. Opposition to abolishing smells was keen, sometimes savage, because it was desperate.[101] The sanitary reformers most often lost the fight; they never succeeded in getting dung buried in trenches. Other measures expected to disinfect the rural house included using lime, opening new casements, and knocking down party walls.[102]

The model projects remain to be considered: the workers' cities of Mulhouse, Brussels, and the rue de Rochechouart in Paris. There are interminable descriptions of the subtle tactics used by their creators and the sanitary reformers, particularly Villermé, to abolish all promiscuity there, to protect the privacy of the family, and to eliminate the erotic encounter in passages and stairways.[103] However, this scholarly, very significant sanitary and moral plan involved only a tiny work force at the time.

More important to the present argument was the attempt to inspect the habitat of the masses. Once again, it was the terrible epidemic of 1832 that prompted new tactics. District commissions were set up when it was announced that the scourge was imminent; their function was to visit every house, detect the causes of insalubriousness, and force the landlord to comply with police rules. These commissions performed their task thoroughly; the one in the Luxembourg district visited 924 properties in less than two months. The prefect Gisquet claimed that he received about ten thousand reports from these organizations.[104]

In England, even before the General Board of Health was established in 1848, the dwellings of the masses had been "harassed by the hygiene police" for some considerable time.[105] Here, the authority was in the hands of local committees. In London, health inspectors arranged visits to houses and sent in a "note specifying which habitation had to be washed, whitewashed, cleared of rubbish, its courtyard or cellars paved, water supplied, drained, ventilated, finally cleaned up in any way whatsoever."[106] A medical practitioner was to judge whether these comments were well founded, and, with his approval, instructions were sent to the landlord, who had to carry them out within a fortnight. In 1853 the inspectors visited 3,147 houses, or 20 percent of the total, in this way, and sent out 1,587 specifications.

The long-called-for French law on unhealthy habitations was finally promulgated on April 13, 1850. Its basis had long been prepared by work carried out in the Conseil de Salubrité since 1846, and it was preceded in Paris by the police ordinance of November 20, 1848. According to the principal craftsman of the new law, the marquis de Vogüé, it tended to establish "a closer patronage" over habitations.[107] The inspection card—the model for which appeared as an appendix to the text of the law—provided for investigation of the condition of latrines and the odors they emitted.[108] Monfalcon and Polinière had cause for congratulation; it was they who had wanted the administration to decide to supervise the dwellings of the poor no less than the animals' cages in the zoological gardens.[109] Passot, for his part, asked that the police inspect the workers' latrines and that they be authorized to make an official report.[110] In fact this law was very rarely enforced; on this point all the sources concur.[111]

10 *Domestic Atmospheres*

*Phobia about
Asphyxia and the
Odor of Heredity* From the mid-eighteenth century onward, domestic architecture, concerned to meet the new requirements of comfort, had been anxious to promote a specialized distribution of living spaces and to designate their particular functions. In the new dwellings, and even more in architects' plans, rooms ceased to be intercommunicating; numerous passages ensured their autonomy. Reception areas tended to be separate from private areas. The capacity to be alone in a well-ventilated space was, according to Claude-Nicolas Ledoux, both a physical and a moral therapeutic imperative.[1]

New sensory requirements directly accompanied this development as early as 1762. Abbé Jacquin had already urged that unpleasant odors in dwellings be combatted and that kitchens be kept clean.[2] He advised against excessive use of water and glazes, release of smoke, and the presence of cats and dogs in the bedroom; he recommended that the "conveniences" be at a distance from it and that the curtains be kept open. His book shows that well before the nineteenth century the location of smells and the tactics of deodorization had their roots in notions of privacy designed to benefit the ruling classes. But after 1832 the seriousness of the scare prompted renewed emphasis on the subject, and the systematic nature of the ensuing proposals underlined the rapid changes taking place in collective psychology.

Once again, what was new tallied with the broad bourgeois aim of both keeping away from and protecting himself against the masses, whom he was in other respects endeavoring to supervise more closely. Deodorization involved a withdrawal into his dwelling, the formation of the private sphere—in short, that process of "domestication" already begun in the eighteenth century that had led Robert Mauzi to write: "the bourgeois is happy only in his home."[3] The implementation of "domestic hygiene" (which tended to become "family hygiene"), like the deployment of bodily hygiene, was no more than the complement to the retreat from public life; together, they engendered a type of habitat dependent upon the medicalization of private space. In the shelter of his dwelling, far from the odor of poverty and its menaces, the bourgeois sought to partake of the fashionable narcissistic pleasures and the subtle bodily messages that now wove a new delicacy into emotional exchanges.

Several important events had occurred since the days of Abbé Jacquin. As a result of Lavoisier's work, it was known that it was not movement of air which purified it. Only renewal could, in a given space, restore the fluid necessary to its proper constitution. Thus in terms of the quantitative and qualitative needs of the organism it was more important to ensure that every inhabitated area was aired rather than ventilated. This need inspired the attempt to define air in terms of cubic norms, modified in accordance with the individuals concerned. Purifyng no longer involved stirring air into movement, but building up reserves of pure air and controlling its flow. The opening of the rue Rambuteau did not lead Dr. Bayard to hope for increased ventilation of public space;[4] it ensured an atmospheric reserve for neighboring dwellings to draw upon.

Inside dwellings, behavior toward smells was more closely concerned with the living breath than in the past. The society of the July Monarchy was very aware of the respiratory phenomena. The new demands regarding quality of air, the revulsion from confined space and the stale odor of rooms, the pervasive threat of tuberculosis that tended to crystallize the terror of death, the phobia of asphyxiation (which was now correctly interpreted and which, in its metaphorical usage, became a stereotype) all revealed the same anxiety about the atmosphere; and the increase in this anxiety was based on scientific authority. Louis Chevalier has shown how the myth of collective asphyxia called for a new interpretation of the town, its space, its buildings and their apertures. He detected the new fear about fog: "vile hotch-potch of all the vapors that are feared," wrote Delphine Gay, "chain of smoke and vapors that marries paving to roofs . . . monstrous fatal union of the breath from chimney and drains."[5] It was a society that, torn between fascination for the private refuge and obsession with "atmospheric captivity," dreamed about "air baths" but snugly shut away its chlorotic daughters and languishing wives.

Sensitivity to "territorial offenses" developed in public as well as in private space;[6] excrement and bodily effluvia were among the ways in which the territory of the "I" could be violated; they became encroachments. Tolerance of the smell created by the proximity of other people diminished. After all, Chaussier had shown that the product of sweating, dispersed into the atmosphere, was eminently subject to putrefaction.

In the private sphere itself, family odor became obtrusive. Around 1840 a new alarm was sounded on this subject. There were hidden dangers in the family, whose virtues had been so highly praised; a specific hygiene was required. This relatively unnoted aspect of pre-Pasteurian attitudes coincided with the (widely studied) emergence of anxieties about morbid heredity and predisposition.

As early as 1844, one of the greatest sanitary reformers of the day, Dr. Michel Lévy, warned of the harmful effects of "the family atmosphere" and "the gaseous detritus of the family."[7] "The family atmosphere" synthesized the individual atmospheres in the dwelling in the same way that the atmosphere of the city represented the sum of social emanations.[8] The imagination of sanitary reformers transposed to private space the threats that had so often been brandished concerning public space. But a specific danger emerged that had no reference to inadequate volume of respirable air or to the absence of collective hygiene; even without any intrusion from the stench of the masses, "the family atmosphere" could be deadly. The accumulated noxiousness of miasmic exhalations, which, by virtue of being related through kinship and heredity, were of the same nature, constituted in itself a morbid menace. The unique mix thus formed impregnated that "habitus vital" of "the domestic atmosphere." As a result of this constant familial "miasmic intercourse," every house had both its own odor and its "specific endemic diseases," kept alive by the mephitism of the walls.[9] Lévy expressed it thus: "We have in mind, not the known effects of the vitiation of the air by congestion, by the escape of flue- or lighting-gas, etc., but the continual exchange of all the influences composing the atmosphere peculiar to several individuals sprung from the same blood, having the same predispositions . . . Cohabitation brings into conflict the personal atmospheres of those who live together; equilibrium results from reciprocal saturation, which strengthens certain morbid predispositions in those who are afflicted with them and develops them in those who were previously exempt from them."[10]

To counteract the harmful effects of "the domestic atmosphere," good family hygiene required creation of an area reserved for the free deployment of individual atmosphere, without risk of reciprocal contamination. The interchange of familial emanations called for a private individual area, in the same way that not so long before the interchange of social emanations had decreed flight from the town or withdrawal into the family dwelling. The revulsion from other people's emanations within the family itself accelerated the process of individualization begun in the mid-eighteenth century. After the successful promotion of the individual bed, the next step was the individual bedroom.

Among the masses such an ambition would, for the moment, have been misplaced; the proletarian family, which sanitary reformers aimed to normalize, was subject to the accumulated effects of family miasmas and had small chance of escaping the threat of disease. The boy's scrofula, the girl's chlorosis were already embedded in the fabric of smells characteristic of the dwelling. The stench of the poor was identified with hereditary decay.

The Demands of the Sanitary Reformers and the New Sensitivity

The new beliefs created norms concerning domestic space and the smell of its atmosphere; they were an incentive to the formulation of new principles. "A habitation is never healthier than when it is alone and isolated," stated Vidalin in 1825 in his *Traité d'hygiène domestique*.[11] It was the house itself that had to be freed from the jumble of social emanations and the miasmic promiscuity of juxtaposed "family atmospheres." This same concern engendered, and then maintained, admiration for English habitations, self-contained, separated from farm, stall, shop, or office. The custom of lodging only one family in a house had triumphed in London, Mille noted.[12]

The other two major imperatives were, as we have already said, proximity to a reserve of pure air and control of the flow of air; they have been the subject of exhaustive studies. But it is worth noting that, well before Michel Foucault, Michelet had revealed the inextricable link between these hygienic imperatives, between the panoptic aim and the wish to improve moral standards. "Ventilation, cleanliness, supervision, three equally impossible things," he wrote concerning the dwellings of the great during the ancien régime, "these infinite labyrinths of corridors, passages, secret staircases, little inner courtyards, finally the roofs, and the flat balustrade roofs provided a thousand hazards."[13]

Investigating the air confined and the odors enclosed in the rooms of the house became the major public health project. Authors of manuals now constantly urged detection of the places where private stenches stagnated. New anxieties generated and governed innumerable descriptions of interiors, although the unexpectedly abundant comments or references supply no evidence that the smell of the atmosphere was getting worse. Nevertheless the descriptions and advice help us to reconstruct the actual smell of the dwellings and the likely source of each odor.

The old threat from the mephitism of the walls, which was manifested more overtly for private than for public space, reflected an old anxiety but also strengthened the new obsession with closet, recess, and corner cupboard, "where air has difficulty in circulating," and darkness encouraged every licentiousness. The small rooms where the children slept rivaled the tiny study of the master of the house in fetidity. Corridors, like staircases, demanded particular attention, because they often impeded proper regulation of the flow. The air in them bred its menaces in stagnation, stench, and darkness or else rushed aimlessly through, causing deadly drafts. As for staircases, without due care they could act as airshafts for the fetid odors of the dwelling. In that case, they sustained the jumble of smells that it was important to destroy—in the same way that meetings between the sexes, which they permitted and which were the first stage in the immorality of the recesses, had to be controlled.

The alcove, "incomplete division of the primitive hut,"[14] was most often fetid and had proved lethal during the cholera epidemic. Its stagnant odor must be shunned; descriptions of this partial refuge of intimacy and pleasure now mentioned only its clammy pestilence. The simple curtain, substitute for the alcove in the modest dwelling, was also to be banned. Particularly close analysis of smells even described the specific character of the air from certain pieces of furniture. The thick atmosphere of cupboards and chests of drawers encouraged proliferation of rats and mice; "moreover, if cupboards are not properly kept, the air that remains in them deteriorates and can, in certain cases, become the source of putrid emanations, which are not without danger."[15]

Featherbeds, wrote Hufeland, became "a veritable potpourri of mephitic emanations, and the person condemned to sleep on such a dunghill for a whole year cannot fail to notice the most deplorable effects."[16] John Sinclair also condemned this repository of unpleasant odors.[17] Charles Londe was more demanding and called for the abolition of pillows and eiderdowns; he denounced the use of too many covers, which activated secretions and encouraged masturbation.[18] The stench of sin germinated in the damp warmth of the body's emanations. Sinclair advised sleepers to adopt nightshirts—and their use did in fact spread—and "to leave collar and sleeves unbuttoned, so that nothing hampers circulation."[19]

The specific odor of the bourgeoisie's rooms is conveyed with a subtlety absent from descriptions of the poor man's hovel. The odor of excrement became less of an obsession—except from the chamber pot, tolerated in bedrooms but considered a problem. Worst was the kitchen. Domestic servants remained trapped within that odor of the kitchen sink which was to offend bourgeois sensibilities for a long time. At the end of the century, the complex odors of the kitchen were still the subject of endless grievance.[20] Effluvia from the ventilation shafts of adjoining closets and from the uncovered rubbish bins warming under the sink mingled with the effluvia of the grubby maid to form an amalgam that very quickly became symbolic of the residual odors of the masses inside the bourgeois dwelling.

Sanitary reformers were very concerned about bedchambers, particularly those of delicate young girls, a concern exemplified by the considerate attitude of César Birotteau, loving creator of Césarine's bedroom. In fact the atmosphere of the bedroom could conceal lethal perfumes. A moneychanger at the Véro gallery succumbed to the mephitism in his chamber;[21] innumerable women and young girls were asphyxiated by flowers while asleep. Progress created new threats, of which fortunately the sense of smell could give warning. The odor of heated metal emitted by stoves (Guy Thuillier pointed out that they were rarely placed in the hoods of fireplaces, which would have reduced the diffusion of carbon monoxide);[22] the coal fumes rising from footwarmers (traditionally blamed for English spleen);[23] the stench of lighting gas, which was replacing or coexisting with the odor of candle grease;[24] emanations from domestic animals—all were simultaneously emphasized and denounced by sanitary reformers. This diatribe increased the popularity of cats, less obtrusive in terms of smell than other pets. Jean-Pierre Chaline has noted that the Rouen bourgeoisie, at a later date it is true, customarily removed footwear from bedrooms in order to preserve sleepers from the unwelcome odor.[25]

Even though individual beds were still not the rule in every hospital and prison—far from it, in fact—the individual room was becoming a common requirement within the lower middle class.[26]

Sanitary reformers therefore set out very early on to define the imperatives that, if respected, would ensure the healthiness and morality of living space. The great Hufeland, whose message quickly gained renown throughout Europe, recommended that not only domestic servants but also flowers and dirty linen be excluded from it; in short, he ordered all possible effluvia to be eliminated. Above all, he demanded that children not sleep and spend the day in the same room. A few years later, Londe summed up clearly the action to be taken: nothing should be kept in the bedroom "that might consume the respirable air or retain around the bed air that has been breathed out. Therefore, no lamps, fires, animals, or flowers. Keep the curtains of the bed or alcove open."[27]

The proponents of sanitary reform demanded twelve to twenty cubic meters of air per hour for this privileged place of respiration. This requirement made vast dimensions necessary, all the more so because windows must not be left open too long. The doctors were anxious to preserve the bedchamber not only from domestic stenches but also from the putrid and immoral stenches of the street.

Did the extreme sensitivity to the smell of the domestic atmosphere of the house remain the prerogative of the sanitary reformers, or were they only reflecting and spreading a new social attitude? An abundance of literary evidence suggests that the second hypothesis is correct. Léonard Pfeiffer discerned an identical sensitivity in Balzac.[28] There are numerous references to the atmosphere of kitchens in the novelist's work (*Un Début dans la vie, Le Père Goriot, Pierre*

Grassou, La Maison Nucingen, Les Comédiens sans le savoir). Balzac showed that he was already aware of the odor of the kitchen sink, the stench of badly cleaned rooms (*Madame de la Chanterie, L'Initié*), the specific odor of offices (*La Maison du chat qui pelote*); in fact the atmosphere of the office, corrupted by emanations from the bachelors who peopled it, quickly became a stereotype.[29] Balzac discerned the effluvia characteristic of unventilated rooms in old townhouses and the odors of the foul-smelling bed, warmed by the body; he denounced "the rank odor exhaled by old tapestries and dusty cupboards" ("L'Elixir de longue vie," in *Les Marana*); on several occasions he analyzed the stench of the death chamber. Fascinated by the coincidence of people and places, Balzac compared the specific odor of rooms and the temperament of the individuals who used them; this method led him in the same direction as the intuition of sanitary reformers, which associated the essence of the dwelling with the specific family atmosphere. From that association, by virtue of this interrelationship, some apartments were held to smell pleasant, particularly in Paris. Their boudoirs and vestibules exhaled an odor of flowers (*Les Employés, La Maison du chat qui pelote*). The chests were of odoriferous wood; the drawers of the commodes that held the young girl's trousseau scented the chamber (*Mémoires de deux jeunes mariées*).

Balzac delighted in reconstructing the smells of the atmosphere of certain semipublic places: pharmacies, ballrooms, concert halls, inns, courtrooms.[30] The boardinghouse caused the greatest repulsion; it "exhales an odour without a name in the language . . . The atmosphere has the stuffiness of rooms which are never ventilated, and a mouldy odour of decay. Its dampness chills you as you breathe it, and permeates your clothing. Smells of all the meals that have been eaten . . . linger in the air. The whole place stinks of the kitchen and the scullery, and it has . . . the reek characteristic of all refuges for the unfortunate. If some way were known to measure the nauseating elements breathed out by every lodger young or old into the air, which is infected by his catarrh or other ailment, it might then be possible to describe this smell adequately."[31] Once again, the novelist's analysis coincided with the deductions of sanitary reformers.

Despite the increased importance attached to the visual, this sensitivity to the specific odor of rooms and furniture persisted until Pasteur's revolutionary discoveries; at times it was very keen. Its development can be followed through studies devoted to Baudelaire, the Goncourt brothers, and, even more, Huysmans; it went beyond the quest for "the soul of the apartment"[32] and overflowed into the frenzied search for harmony between rooms and personal mood (*Stimmung*). The fusty odor of Uncle Adolphe's room or of the shooting lodge in the forest produced revelations as powerful as the little madeleine or the paving stones of the hotel Guermantes.[33] Half a century later, Bachelard set out to analyze the sensory composition of private space.[34] The important task here is to date the genesis and parentage of this new sensitivity, whose privileging of the revelations of the sense of smell has a long history behind it.

The sudden awareness of the characteristic smells of the rooms that composed the private home created a desire to promote the specific smell of individual rooms and thus abolish the offensive mixture of the family atmosphere; it was an incentive to check as far as possible the formation of the potpourri of domestic odors. Like the promiscuity to which it testified, confusion of smells had become obscene. Apart from eliminating the confined air of the recess, the only way to get rid of troublesome smells and to reserve private space for the delicate effluvia of intimacy was to sort them out and contain the strongest smells in the appropriate places. A new intolerance proscribed the mixing of organic odors and subtle perfumes; preventing such confusion would be the function of the modern kitchen, dressing room, and bathroom.

This demarcation of the intimate spaces of interior monologue liberated the olfactory possibilities of both bedchamber and salon; they sanctioned the emergence of an aesthetic of olfaction within private space. A science of scents designed to embellish places of intimacy accompanied the cautious advance of perfumery. Their progression was governed by the same concern to produce increasingly subtle individual messages, and the same desire to reveal and emphasize the person. Because they obeyed the same imperatives, the skillful arrangement of the background of smells in the boudoir came to be aesthetically inseparable from the odor of the woman to whom one came there for inspiration.[35]

Symbolic of the process was the growing concern to deodorize the individual bedroom, which came to be regarded as the place of the smells of intimacy par excellence. Within this refuge, separated lovers could indulge in solitary respiration of the beloved's perfumes.[36] Odors helped make the chamber the mirror of the soul.

The skillfully delicate atmosphere of this refuge for secret tears and pleasures tended to replace the sensual animality of the alcove.

Actions and Norms

At the end of the eighteenth century, ships, barracks, prisons, and hospitals served as the laboratories where the techniques of ventilation and disinfection were formulated. In the following century these were the places where the hygienic activities appropriate to the household were specified and the norms of ventilation defined according to the new scientific imperatives.

Accompany Howard, not to a prison this time, but to the Venice lazaretto. Follow him into the room where the employees purified merchandise. Packages were moved about, shaken up, turned inside out; sheets were unfolded, shaken, sometimes spread out on nets; furs were shaken, skins beaten, every object exposed to the air.[37]

Now listen to Charles Londe (1827) on the subject of housework in the chamber: "It is necessary to shake the sheets, covers, mattresses, and bolsters every day and, while this is being done, to set up a draft in the apartment by opening windows facing each other." In addition, it was imperative to beat the mattresses at least once a year in order "to rid them of putrescent animal substances."[38] John Sinclair also set out to codify the activities of everyday housekeeping: to drive "all harmful vapors" from the bed, one must "open the windows of the room and freely expose sheets, covers, and curtains to fresh air."[39]

Airing, beating, lifting, shifting, and hunting down with a broom the menace of recesses became the dominant forms of housecleaning. It was not so much a matter of contending with dust as of ridding the furniture and the various rooms of their vitiated air, of eradicating stenches, of anticipating putrefaction.[40] Dust, like spiders' webs, only indicated lack of ventilation. It naturally provoked numerous scientific works, but only in connection with attempts to discover the possible presence of putrid substances in it.[41] And if Forget advocated constant use of brooms on ships, it was because he expected them to eliminate the waste and detritus that might accumulate in inaccessible places.[42] We must be careful to avoid the anachronism of identifying nineteenth-century housekeeping practices with the dust neurosis created by Pasteur's revolution.

The major undertaking was still the determination of spatial norms to meet respiratory needs. Arbuthnot had already tried to measure the minimum volume of air below which the individual was condemned to die. Howard peremptorily stated that the prisoner's cell had to be ten feet long, ten high, and eight wide, but without bothering to justify this injunction.[43] Tenon asserted that the height of hospital rooms had to be modified according to the nature of the disease; the feverish patient demanded more air than the convalescent.[44] In 1786 Lavoisier suggested a cubic norm.[45]

In the next century, following progress in the analysis of confined air, scientists set out to make this "adjustment of space and the organic body" more precise;[46] it was another of those tasks that began to look Sisyphean as the years went by. Despite the imprecise measurements, Leblanc and Péclet finally reached agreement. They estimated that the individual demanded between six and ten cubic meters of air an hour.[47] This was the quantity that domestic hygiene experts accepted;[48] however, they were cautious, and considered— without regard to exact scientific measurement—that the sleeper shut in his chamber should be provided with a double volume. By induction Monfalcon and Polinière later fixed on twenty cubic meters an hour for the needs of the horse in the stable.[49]

It seemed that an optimum space could be determined by reference to this respiratory norm. An arbitrary calculation was made: the volume of air put at the disposal of an individual confined in a given space varied according to the strength of the flow; in addition, needs differed in accordance with subject, temperature, and degree of humidity. Nonetheless Péclet concluded that, after a deduction for the volume of solid bodies, the optimum volume for a hospital room with thirty beds was 1,335 cubic meters. Piorry, then Monfalcon, applied this norm to domestic space.

The administration was quick to record the sanitary reformers' calculations. A police ordinance of April 20, 1848, made it obligatory to supply every individual with an area of 14 cubic meters. In the same year, the Commission des Logements Insalubres, formed by the Paris Conseil d'Hygiène, suggested a volume of 13 cubic meters per individual per chamber. Moreover, gas physics confirmed the validity of sanitary reformers' advice in the previous century. The height of the ceiling of the chamber should be no less than 3 to 3.50 meters, wrote Piorry; otherwise the head would be "in the region where the lightest, most insalubrious gases are borne."[50]

All these calculations, however, remained largely theoretical. No one in France has yet cataloged the achievements produced by these tactics. Nevertheless, in some towns the design of the traditional dwelling was recast in the course of the century as a result of three factors: the sanitary reformers' injunctions; the desire for comfort, which stimulated separation of spatial functions, differentiation of service areas, reception, and family intimacy;[51] and the quest for profit, which provided an incentive to increase the area of the house available for rent. For example, in the typical Lille house described by A. de Foville in 1894, everything was arranged so as to exclude troublesome effluvia: "kitchen, scullery, and latrines are relegated to an outbuilding, and the unwholesome odors that emanate from it are lost in the yard and garden, without penetrating the dwelling."[52] An identical revolution occurred in Tours; a new, unaesthetic building took up part of the tiny "individual" garden; it housed the kitchen on the ground floor, the washrooms and laundry on the first and second floors. These few prototypes worked out at the beginning of the century provide rich models for the future, although their significance is likely to be underrated by historians unalive to the extent that these minor and somewhat marginal experiments presaged later developments.

The initiative in household design belonged to the United Kingdom. The English innovations fascinated the French even while they rejected them. In Britain "no one disputed this truth: unpleasant odor in the dwelling . . . indicates an outrage to public health." English tactics, as summed up by Mille, were based on the adoption of three principles: "water at full pressure and a free tap," notably in kitchen and water closets; disposal of sewage through drains; and the adoption of the new machinery of comfort, remarkably well studied by François Béguin.[53] In the English plan, in fact, control of the water flow accompanied control of the circulation of the air and the automatic expulsion of rubbish.

Around the middle of the century, 300,000 dwellings in London were supplied with water. In Glasgow "a water closet, a hot bath, and a shower bath are found on every floor in well-to-do houses everywhere." In some medium-sized towns, water supply and mains drainage were installed at the same time, financed by a municipal tax. Total cleansing was achieved immediately. Of 1,100 houses in Rugby, "700 to 750 have been connected [to water] and have at least two taps, one in the kitchen, the other in the water closet."[54]

The same progress was recorded in Croydon, Warwick, and Dover.

In a few decades a gap opened up between the British Isles and the Continent.[55] The relative indifference shown by the French to cleanliness, their rejection of water, their long tolerance of strong bodily odors, and their continued privatization of excrement and rubbish cannot be explained solely by a secret distrust of innovation, by relative poverty, or by slow urbanization. It was the collective attitude toward the body, the organic functions, and the sensory messages that governed behavior patterns. It is regrettable that historians have given scant attention to this somatic culture.

Given France's rejection of mains drainage (except at Lyons) and its slowness in installing water supply and in implementing the machinery of comfort, model achievements there usually concerned only ventilation and the new arrangement of domestic space. A crucial feature of French policy was the installing of rooms in suites; according to Lion Murard and Patrick Zylberman, the idea dated from 1827: it is expressed in d'Arcet's *Description d'une salle de bains,* published that year.[56] Seventeen years later Piorry summarized the new imperatives of chemists and engineers: "a good kitchen should be vast, very high, flagged, well cleaned, ventilated close to ceiling and floor"; a hood should be installed "communicating with that of the main hearth, and its opening calculated in such a way as to form a draft and to carry away the exhalations from the coal"; a grooved cloche would dam up the odor of the sink.[57]

The scarcity of water and the absence of an evacuation system challenged the ingenuity of sanitary reformers charged with devising latrines. The importance of the issue was not underestimated. "The latrine," concluded Grassi, chairman of the commission instructed to study this problem in 1858, "must be the cleanest place."[58] By a curious turnabout—in some ways reminiscent of Parent-Duchâtelet's attempt to make the sewer-cleaner a model of the morally improved worker—cleaning up the privies was expected to produce a chain reaction resulting in the deodorization of private space. Like the latrines in the asylums and the dwellings of the masses, the bourgeois privy became the favored place for apprenticeship in the disciplines of hygiene. It was Grassi again who stipulated that "there must be a niche or some sort of obstacle over it [the seat] to prevent visitors from climbing up on it and taking up any position other than that indicated by its name." Keeping the privies clean was only a "matter of supervision and discipline."[59]

The enforcement of discipline in defecation in the schools prepared for the diffusion of attitudes into the private domain; it generated an abundance of literature.[60] Inspectors and sanitary reformers determined the norms, chose the furniture, conducted manifold experiments, cited examples of headmasters who had obtained obedience. For example, the head of the boys' school in the rue de la Réunion managed within a few days to inculcate in his pupils the habit of sitting, and no longer climbing, on the seats.[61] One imperative was endlessly repeated: the master must be able to observe the ceiling and floor of the cubicles from his chair.[62] Discipline was even more severe in young girls' boarding schools; the educators firmly recommended retention to their pupils; by control of her physiological needs, a gentlewoman had to prove that she was able to resist all the urges of the body.[63]

Sanitary reformers were not content to advocate ingenious machinery. They now aimed at turning the lavatory into a real room—a development that could only increase in importance in the private house. Its astonishing and significant rise was accompanied by increasingly luxurious decoration, which reached its peak in Victorian England and in Wilhelmine Germany.[64] Grassi described the model water closet: the bowl, which was equipped with a siphon, took the form of an earthenware or glazed terra cotta funnel. The seat and its lid were, like the floor, of polished oak. A urinal was situated close by so that the chamber pots could be emptied without spreading the pungent odor of stale urine through the house. The siphon or artificial suction ensured deodorization; as a last resort, a blowoff pipe fixed to the arch of the hole ensured the evacuation of stenches. Multiple seats, as well as Turkish-type holes separated from each other by a simple handrail, were to be forbidden.[65] The overriding goal was to eliminate the old promiscuity in defecation and the jumble of excremental odors, now held to be intolerable.

In practice such conveniences were still the prerogative of a small minority. In the provinces it was still common practice, even among the bourgeoisie, to tip excrement onto rubbish dumps, even into roadways. In Le Havre in 1849, only new houses built by the rich had cesspools;[66] denounced for nearly a century in Paris, they seemed like progress here. In the Place Manigne in Limoges, a hundred meters from the town hall, overflow drains were still in use at the beginning of the twentieth century.

Washrooms appeared later than modern water closets; when they did become quite widespread, toward the end of the century, they were usually only "unheated recesses furnished with a bowl and a water jug";[67] Jean-Pierre Chaline observed this convenience among the Rouen bourgeoisie, although he forgot to mention the bidet, sometimes pushed under the table.[68] The steam from the bowls, the vapors of sponges mingled with the perfumes of essences, produced a heavy atmosphere in these exiguous places. At least the odors of soap were no longer dispersed to the bedroom. The belated appearance of these washrooms was an important landmark in the long process that allotted specific functions to the places of intimacy, and was indisputably the major event in the history of domestic space in the nineteenth century.

Deodorization of the washing area came only after the even more belated spread of the bathroom—which should not be identified with the practice of taking a bath. For a very long time, the bathroom was found only in the dwellings of the rich, tourist hotels, and luxury brothels.[69] Alfred Picard noted in 1900 that in Paris only apartments with very high rents were furnished with bathrooms.[70] For a long time the nudity of moving bodies, the total freedom of the gestures of the toilette, and the cozy intimacy sheltered from all intrusion conferred on the place a perfume of licentiousness, further emphasized by the figures of Leda that were often sculpted in the bathroom fittings.

The rare nineteenth-century bathrooms, equipped with heavy furniture and protected by warm hangings, tended to be spacious; we have seen why. Sanitary reformers recommended the use of metal bathtubs; marble was too cold. They advised paneling the partitions to protect against the mephitism of the walls, and above all using solid partitions to shelter the bedroom from condensation and moist odors. At the beginning of the twentieth century, with the exclusion of traditional furniture and the adoption of sanitary equipment fixed by plumbing in a rigid arrangement, the bathroom became deodorized. Later the use of "clean and decent"[71] geometric space came to underwrite the idea of the bathroom area as a sensually neutral and innocent space.[72]

11 *The Perfumes*
of Intimacy

THE NEW CONTROL of odors that accompanied the increased privacy inside bourgeois dwellings permitted a skillful change in the way women presented themselves. A subtle calculation of bodily messages led to both a reduction in the strength of olfactory signals and an increase in the value assigned them. Because, in the name of decency, women's bodies were now less on show, the importance of the sense of smell increased astonishingly. "The woman's atmosphere" became the mysterious element in her sex appeal. However, exaltation of the young girl's virginity and new perceptions of the wife, her role and her virtues, continued to forbid any open advances. To arouse desire without betraying modesty was the basic rule of the game of love. Olfaction played a crucial role in the refinement of the game, and it turned primarily on the new alliance between woman and flower.

"Persistent
Cleanliness"

New medical arguments were invoked to justify practices aimed at eliminating putrid filth so as to diminish the risks of infection.[1] Since the time when Lavoisier and Séguin had precisely measured the products of cutaneous perspiration,[2] there had been increased concern that perspiration not be impeded. Broussais's physiological approach to medicine called for greater attention to the hygiene of the secretory organs, which ensured that the body was freed from impurities. Medical theory indicated the "geography" of the body that governed the ritual of the toilette. The key was to supervise cleanliness of hands, feet, armpits, groin, and genital organs. The importance that Broussais accorded to the concept of irritation strengthened the ban on oxy-metallic cosmetics.[3] Sensualism, which still had considerable influence though it was now questioned, enjoined sensitivity and lightness of touch by means of a scrupulous toilette.[4]

The canons of bodily aesthetics urged the most scrupulous hygiene. Cosmetics were governed by the aristocratic ideal of pearly skin, through which the blue blood could be seen pulsing. For nearly a century the supreme reference points remained the brilliant whiteness of the lily and of the Pompadour's complexion;[5] the aesthetic code decreed that all visible parts of the body be washed—as it also prescribed a sedentary life, the cool shade of trees, and the protection of gloves for "soft, white, firm, plump hands."[6]

Removing the dirt from the poor was equivalent to increasing their wisdom; convincing the bourgeois of the need to wash was to prepare him to exercise the virtues of his class. Cleanliness was the tenth of Franklin's thirteen principles of wisdom, coming just before moral balance and chastity.[7] "Hygiene, which maintains health, which nurtures the mind with habits of order, purity, and moderation, is for that reason alone the soul of beauty; because this precious advantage depends more than anything else on the freshness of a healthy body, the influence of a pure soul."[8] Vidalin put his finger on the unexpected link between economy and cleanliness.[9] Cleanliness, in its widest sense, reduced waste of food and clothing and facilitated the identification, control, and even possibly the salvage of waste; it became another method of fighting loss.[10] From this point of view, the best preventive measures consisted in learning not to get dirty, to avoid contact with the putrid, and to get rid of all excreta on the skin.

The stress on the requirements of modesty, as we know, both favored and restrained the practice of bodily hygiene. Olfaction was caught up in the network of intertwining prohibitions. Richard Sennett has cited the physiological and psychic disorders that fear of farting in public caused the Victorian bourgeoisie.[11] Although contemporary hygiene manuals scarcely alluded to this retention, they evinced a quite new form of delicacy with regard to smells. Do not require your servant to do anything repugnant to her senses, the comtesse de Bradi advised in 1838; except in case of illness, "do not have your shoes taken off."[12]

Nevertheless the progress of bodily hygiene encountered numerous setbacks; the most important was the slowness with which houses were equipped, authorized by doctors' persistent distrust of immoderate use of water, as is evidenced by the long list of prohibitions and precautions that larded public health discourse. Menstrual frequency still ruled the cycle of bathing. Few experts advised taking more than one bath a month. Hufeland was notoriously bold in prescribing a weekly regimen, and Friedlander even more so for permitting children to take baths two or three times a week, although he still denounced immoderate use of water.[13]

Plunging into water involved a calculated risk. It was important that the duration, frequency, and temperature of baths be adapted to sex, age, temperament, state of health, and season. Baths were thought to exert a profound effect on the whole organism, because they were not an everyday event. Mental specialists and sometimes even moralists pinned their hopes on them;[14] gynecologists feared them. Delacoux noted that the courtesan owed her infertility to her excessive preoccupation with toilette. According to him, numerous women had been deprived of the joys of motherhood by this "indiscreet attention."[15] Even more serious, baths endangered beauty; women who overused them "were generally pale and their fullness of figure owed more to fleshiness than to the bloom of the skin."[16] Young girls who bathed too much risked debility.

Tourtelle advised against immersion after meals, during times of weakness, and, of course, during menstruation. Rostan recommended that the bather wet his head in order to prevent congestion of the brain.[17] One must get out of the water at the occurrence of the second shiver, dry promptly, and then stretch out on a bench for a few minutes to rest from the fatigues of the bath, without running the risk of making the chamber damp.

Until the triumph of the shower, which shortened the time spent in bathing and rendered self-satisfaction harmless, baths aroused suspicion. The ban on nudity acted against the spread of baths. Drying the genitals posed a problem. "Close your eyes," Madame Celnart ordered her readers, "until you have completed the operation."[18]

Water could become an indiscreet mirror. Dr. Marie de Saint-Ursin described the young girl's confusion in the bath: "Inexperience descends, blushing, into the crystal of the waves, meets the image of its new treasures there, and blushes even more."[19] In affected phrases, the author confirmed the synchronism established between female puberty and initiation into the practices of bodily hygiene.[20] "Bathe, if you are ordered," concluded the comtesse de Bradi; "otherwise take only one bath a month at the most. There is an indefinable element of idleness and flaccidity that ill becomes a girl in the taste for settling down on the bottom of a bathtub in this way."[21]

These attitudes account for the obvious disparity that grew up between the volume of discourse and the scantiness of practice.[22] Bathing still required therapeutic justification. It is therefore not surprising that the ritual was complicated. Transporting the water and filling and emptying the tub, bucket, or metal bath entered, like laundry or seasonal housework, into the timetable of major domestic rites.

Accordingly, the major innovation was an increase in partial baths, as shown by the spread, still limited it is true, of footbaths, handbaths, hipbaths, and halfbaths. The concern to avoid getting dirty, the new frequency of ablutions, and the stress on the specific requirements of washing shaped the bourgeoisie's apprenticeship in hygienic practices. The physiology of excretion, which appeared even more important in the light of Broussais's theories, governed the fragmented ritual of the toilette in the same way that it obsessed both the actual practices and utopian fantasies of municipal policy. The same rationale underlay both the insistent hygiene advocated for the bourgeois body and the ceaseless evacuation of urban waste that was the sanitary reformers' aim: the abolition of the threat from excrement, no longer so much determined by the risk of infection as by the risk of congestion.

A proliferation of lotions accompanied the increase in ablutions. This was the result not only of a process of substitution encouraged by the disrepute of perfumes, but also of the alliance between lotions and friction, highly recommended for its energizing properties. The other activities of the toilette ritual can be quickly listed. The fashion for plastering pomade on hair had disappeared.[23] Hair hygiene consisted of periodic untangling with a fine comb, brushing, and plaiting before bed. The Salernitan prohibition remained: the head was not washed. Madame Celnart recommended rubbing hair with a dry cloth to remove dust; at the most, and then with caution, the fashionable lady could use a soapy lotion applied with a sponge.[24] The practice of shampooing developed only during the Third Republic—fortunately, because until then the strong scents of her hair remained one of woman's strongest trumps; she had, after all, been forbidden to use too much perfume.

Mouth hygiene was becoming more specific; Londe advised daily brushing of all the teeth to deodorize breath, and not, as was most often the practice, only the front ones.[25] Madame Celnart prescribed the use of aromatized powders.[26]

Fresh bodily odor depended even more on the quality and cleanliness of underwear than on scrupulous hygienic practices.[27] Development in this area too moved at an accelerated pace. Sanitary reformers endeavored to institute weekly changing of underwear. The new timetable for changing clothes[28] and the new appreciation of the pleasant odor of clean linen were incentives to perfume washtubs, chests, and drawers. These practices spread long before those of bodily hygiene proper.[29]

In fact the new behavior patterns were accepted only slowly even by the bourgeoisie, as the rareness of bathrooms attests. The bidet was popularized only at the very end of the century.[30] The use of the tub, imported from England, long remained a sign of snobbishness. In 1900 a Parisian bourgeoise of good social standing was still quite happy with periodic footbaths.[31] Inventories suggest that contemporary doctors may have possessed a fair number of halfbaths, but this was because they formed an avant-garde responsible for encouraging hygiene.[32]

For the time being there could be no question of obliging the masses to follow a ritual that the elite was still neglecting. Consequently, they remained condemned to steep in their oily, stinking filth, unless they braved the putrid and immoral promiscuity of public baths. Practices of bodily hygiene became commonplace in the Nivernais only after 1930, according to Guy Thuillier.[33] Until then, the apprenticeship in cleanliness carried out at school, barracks, and youth organizations aimed at scarcely more than external appearances; the battle over the use of the comb, the ritual cleanliness inspections carried out by schoolmasters, and the advice dispensed by Madame Fouillée in *Le Tour de la France par deux enfants* were clear proof of this.[34]

Nevertheless, several segments of the population were already confronted with the norms formulated for the bourgeoisie. Prisons rather than boarding schools were serving as the laboratories for personal hygiene. As early as 1820, Villermé ordered that convicts comb their hair before washing their faces every morning, and that they wash their hands several times a day and their feet every week.[35] He advocated a weekly review of cleanliness; he wanted new arrivals to be bathed and the administration to insist on short hair. Sanitary reformers asked no more of school children a century later.

Wet nurses for the bourgeois newborn were compelled to follow norms that were probably stricter than those prevailing in the nurseling's family. Doctors advised that they be bathed once a month and forced to wash their mouths, breasts, and genitals every day.[36] It is hard to assess the influence these women exerted when they returned to the village.

In the countryside, water supply had to be regulated before the traditionally dirty classes, whose dip in the river was their sole practice of cleanliness, could enter the prevailing value system of sanitary reformers' precepts. A major advance, still very little studied, took place around the middle of the century. The example of Minot may well be representative. A complex system of watering places, cisterns, beechwood buckets, washhouses, and fountains—in fact a new aquatic architecture—was constructed in the commune; in 1875 the circulation of water was mastered and organized within the village itself.[37] While female social life ceased to be focused on the well and moved to new working areas, the complex tactics of domestic and bodily hygiene, refined in the town, began timidly to spread. In this environment too, control of the water supply would permit new patterns of everyday life.

Smell, Perfume, and New Perceptions of Elegance

Under the July Monarchy, the elegant gentleman had stopped using perfume—unless he played the dandy or practiced "antiphysical" love. At the most, his person emitted a vague smell of tobacco, which he had to try to keep from the ladies.[38] For him, the time for ostentation was past, as historians of fashion and costume have clearly demonstrated. The new code of male elegance was subtle; there was no longer a place for nuances of smell. It was precisely the absence of strong odor, evidence of careful hygiene, that was regarded as the decisive criterion of good taste. The symbolic, barely perceptible, odor of cleanliness that emanated from linen characterized the deodorized bourgeois.

The wife, on the other hand, had become the man's standard-bearer and "the ceremonial consumer of goods which he produces."[39] She was invested with the function of representing the position and wealth of father or husband. Silken draperies, bright colors, ostentatious luxury were henceforth her preserve; they testified to extravagance that placed her above all suspicion of work.

In terms of scents, the code of elegance became more refined. Up to the end of the century, the range of permitted scents remained very small; despite short-term oscillations in fashion, good society respected the aesthetics established at the court of Marie Antoinette. Particularly during the July Monarchy, the hygiene advocated by doctors in matters of smell was an invitation to keep faith with the values of delicacy, to be content with the sweet perfumes of nature, and to eschew the heavy animal scents of musk, ambergris, and civet.[40]

A new use of cosmetics accompanied this quest for delicacy. It caused beauty to be identified with "elegant cleanliness."[41] Paint (red and white) and powders were abandoned and pomades used in moderation.[42] Tourtelle gave a perfect summary of the new requirements of fashion and hygiene: "True cosmetics are aqueous lotions for cleanliness, and unguents, which can be used to cleanse and soften the skin, like emulsive substances, fresh oil, whale soap, butter, cocoa butter, soap, almond paste," and, most of all, he added, "no metallic oxide."[43] What was important was to tear off mask and plaster, air the skin, open the pores, and thus allow the woman's atmosphere to be diffused.

All observers provided evidence of the decline in perfumes; the professional perfumers deplored the trend, especially Eugène Rimmel, one of the greatest.[44] Women of fashion had forsaken homemade perfumes, and the decline of the brew of odors that had hitherto formed a subtle sensory apprenticeship cannot be overstressed.[45] Bath perfumery had practically disappeared, Louis Claye noted in 1860.[46] Once powder had been abandoned, the practice of perfuming hair was the subject of protracted debate; it seems that only the most coquettish dared be bold in this respect.[47]

Good taste forbade the young girl to use perfume; this indiscreet solicitation might reveal her ambitions for marriage too crudely. It might also compromise one of her main trumps. There was absolutely no need to mask, however lightly, the effluvia that emanated from the slender body, the nature of its odor as yet unspoiled by male sperm: "the tender odor of marjoram that the virgin exhales is sweeter, more intoxicating than all the perfumes of Arabia."[48]

In no circumstances should real perfume be applied to the skin. Only aromatic toilet waters—distilled rose, plantain, bean, or strawberry waters—and eaux de cologne were permissible.[49] Keeping perfume at a distance from the body was more important than ever. This increased severity was accompanied by a reduction in the range and volume of objects carrying perfume. Whereas it might be in good taste to impregnate linen with delicate scents from cupboards, it was no longer so to perfume the towels used for the toilette. Sweet odor was concentrated on handkerchiefs and one or two accessories: fans, the lace surrounding the tiny bouquet carried to the ball, and, for the most sensual, gloves, mittens, and slippers.[50]

But there were pleasing compensations for these renewed prohibitions. Perfume impregnated familiar objects, and their odor reverberated and attested female charms from afar. Its function was to form a scented casket, both to allow perception of the woman's atmosphere and to enhance it—in short, to reconcile the irreconcilable. Distance, in the event, assisted seduction; eroticism benefited from this immodest modesty.

This complex plan dictated and justified the abandonment of animal perfumes and the rise of the fashion for floral odors; the latter, without competing with the odors of the flesh, echoed the traditionally mysterious collusion between woman and flower. "It is on the perfumes of nature at the first rays of the sun that the sense of smell must be exercised," decreed Londe in 1838.[51] The comtesse de Bradi added, in a conciliatory vein: "I have forbidden you prepared perfumes; but those spread by natural flowers seem to me very permissible, when they are in no way disturbing."[52] The proportions as much as the nature of the products set the seal on elegance. The list of acceptable perfumes and toilet waters remained essentially unchanged until the middle of the Second Empire. When perfumers set to work around 1860 to refine their products, their basic range of preparations for handkerchiefs remained very small. According to Rimmel, it comprised six elementary odors: rose, jasmine, orange blossom, acacia, violet, and tuberose.[53] The perfumer invented bouquets by combining these six odoriferous bases. To produce pomades, he could also use jonquil, narcissus, mignonette, lilac, hawthorn, and syringa. The Parisian masters, Debay stated in 1861, "have banished strong and intoxicating odors that are harmful to the nerves . . . and offer only innocent perfumes."[54]

Contemporaries justified this timid attitude to smells. Doctors endlessly repeated the old arguments refined at the end of the eighteenth century to discredit animal perfumes, thereafter regarded as putrid substances. They congratulated themselves on their virtual disappearance. The acknowledged need for a good hygiene of respiration created even greater mistrust. There was an increased fear of the ravages made by animal perfumes on the psyche of those who wore them; it accompanied the development in psychiatry. "Misuse of perfumes gives birth to all the neuroses," wrote Dr. Rostan as early as 1826. "Hysteria, hypochondria, and melancholia are its most usual effects."[55] The danger was particularly deadly for chlorotic girls, victims, like pregnant women, of aberrations of the sense of smell (parosmia), even of cacosmia, in which "the odor of burnt horn or other more or less evil-smelling odors," noted Dr. Obry in his thesis on the subject, "is not only tolerated but even avidly sought."[56] This result by itself would be enough to warn young girls, all susceptible to chlorosis, against use of perfumes.

The ambiguous theme of the immorality of penetrating, stifling scents is implicit in medical discourse at those points where it explicitly warns its female readers. Early in Pasteur's revolution the diatribe took on new violence. The charm of perfumes, the search for "base sensations," symptoms of a "soft, lax" education, increased nervous irritability, led to "feminism," and encouraged debauchery. Tardieu's "sniffers" joined the lengthening list of unfortunate "perverts."[57] Tonic and disinfectant lotions were the order of the day.

These psychiatric tactics were more moralizing than when the diatribe against heavy perfumes primarily reflected fear of infection. They helped to relaunch osphresiology, which had been somewhat dormant, like the use of perfumes, since Hippolyte Cloquet's publication of his heavy tomes. Current experimental psychology in particular displayed a new interest in olfactory sensation.[58]

But the survival of the fashion for natural perfumes and the persistent ostracism of those who wore provocative animal perfume clearly had a much larger significance. These refined behavior patterns in relation to smells are a source of information on social psychology. I would like to indicate a few, still dimly lit, paths, without suggesting that any one of them is more important than any other.

"The bourgeois does not employ his wealth to make a show," wrote Robert Mauzi; he needed it in order to exist.[59] This view would be enough to explain the hostility displayed toward perfume: it was a symbol of waste; its dispersion was evidence of intolerable loss. But this argument does not seem to fit the nineteenth-century bourgeois. He was no longer solely that man of duty, the moralist, opposed to enjoyment and even all sensuality, whom Werner Sombart depicted. Obsessed by the desire to legitimate his position, the bourgeois now envied and tried to emulate aristocratic nonchalance. As the years went by, he ceased to give the impression of being socially inhibited. Where ostentation was concerned, it could even be said that he went too far; the fashions of the Chaussée d'Antin rapidly outstripped in magnificence the discreet charm of the boulevard Saint-Germain. Until well into the July Monarchy, it was among the elite that the criteria of good manners were determined. Philippe Perrot, among others, has clearly demonstrated how the new attitudes to simplicity adopted in this milieu governed the code of elegance. From the Restoration on, the hierarchies became more refined, the symbols more complicated; unforeseen divisions became apparent. While the new practices of cleanliness distinguished rich from poor, criteria imperceptible to outsiders fragmented the world of wealth. The cultivated delicacy of the messages of smells undoubtedly formed part of the complex tactics of this set of distinctions.[60]

In addition, in that environment where refinements multiplied, attachment to floral scents and contempt for animal perfumes might be interpreted as a return to the fashions of the last years of the ancien régime and a rejection of exaggerated tastes—if not of the dandies, symbols of counterrevolution, at least of the ultrafashionable ladies and upstarts of the Directory. They were evidence of the rejection of the forms of ostentatious dress at the imperial court. Nevertheless, there seem to be more reliable explanatory factors than those based on interpretation of fashion.

The arbiters of nineteenth-century morality placed modesty above all other feminine virtues, and the prohibition on cosmetics, as well as on indiscreet perfume, was part of a complex system of visual, moral, and aesthetic perceptions. "Unaffected cleanliness, the natural elegance and graces of body and mind, *sprightliness* and modesty are the most powerful cosmetics."[61] The thick vapors of impregnated flesh, heavy scents, and musky powders were for the courtesan's boudoir or even the brothel salon. The obverse model furnished by the venal woman facilitated the definition of true elegance.

The increasingly prevalent symbolism of the natural, sweetly perfumed woman-flower disclosed a firm wish to control emotions. Delicate scents set the seal on the image of a diaphanous body that, it
was hoped, simply reflected the soul. These were ambitious tactics
that attempted to render harmless the threat of animality and to
imbue the woman's impulses with wisdom. She should be rose or
violet or lily—certainly not feline or musky;[62] floral images supplanted those borrowed from the carnivorous cycle. Even within the
vegetable kingdom, imagery was drawn from innocent rural or kitchen-
garden flora, not yet from the disturbing exoticism of lianas or poisonous corollas. A persistent and fascinating attempt to sanctify underlay the fragile symbolism that proliferated around the young girl
(a subject that has not been studied for far too long for fear of
derision). Let the woman deck herself with flowers as she decorated
the altar to the Virgin with flowers, let her adorn her body as she
adorned the wayside altars on Corpus Christi, let her abundant virtues perfume her life like the flowers entwined on the mementos of
first communion, and there would be no possibility of ravaging animality. The contemporary homilies on chastity were probably as
significant—probably even greater in this context—as medical discourse.

*The Calculus of
Bodily Messages*

Modesty seems to have become an erotic science. The allusive coldness, the delicate invitation, the admission of imperceptible
anxieties, the blushes, the constant reference
to the thoughtlessness of the misconduct about
which it was fitting to know nothing—all appear to have been skillful
sexual tactics that integrated the subtlety of the messages of smell.[63]
Were not the natural effluvia from the virgin's body and her airy
toilette perceived as the most erotic traps? Debay summarized this
erotic science of modesty in a few words: "Perfumes should be applied sparingly so that people will keenly desire to smell them."[64] A
new role devolved on olfaction within the framework of these global
tactics. The perfumed invitation was more delicate, less obvious, and
less coarse, perhaps more disturbing, than the charms of nudity; it
was more in tune with the ambiguous wish to seduce. It had the
additional advantage of preserving the appearance of innocence. The
message of love that emanated from the sweetly perfumed body

could no more implicate modesty than could the involuntary curves, hidden but revealed, even accentuated, by the fabric of the bodice.

Skillful perceptions of the sense of smell accompanied—or did they even precede?—fashionable practice. Never again would so much attention be brought to bear on the specific character of individual odors. The expert Barruel claimed to have discovered the scientific method of recognizing them; he offered his discoveries on the odor of blood to the police.[65] Before fingerprints were in use, he suggested the use of "smell" prints—an unknown page in the history of identification.

For the time being, doctors unanimously refused to acknowledge the erotic role played by the odor of sexual secretions in man. This odor whetted the generative instinct in animals, stated Rostan, but "the same is not true of human species."[66] Among humans, explained Londe, the erotic function devolved on the sense of touch;[67] caresses alone were exciting. In animals, wrote Hippolyte Cloquet, the sense of smell was the sense of violent appetites; in man, of sweet sensations.[68] Consequently, blacks, who had remained nearer to the beast, as the anthropologists Blumenbach and Soemmerring had apparently proved, were more sensitive to the sexual power of odors.[69]

Placed in the center of the domestic scene, woman became its stage director; within the bounds of what her modesty permitted, she skillfully calculated the erotic possibilities provided by the framework of her life, transforming it into a veritable forest of symbols. These images are more clearly apparent in the interiors of homes than anywhere else. Even Balzac conceded this in the *Physiologie du mariage:* as long as they were not provocative, vegetable perfumes could enhance the atmosphere of chamber and boudoir. On the other hand, musk, lilies, and tuberoses were forbidden and roses mistrusted.

Perfume-pans were still an essential part of the fashionable young girl's equipment.[70] Fragrant pastilles had not disappeared, but they tended to be reserved for sickrooms.[71] The new fashion was for the perfumed candle, its innocence guaranteed by the fact that it was useful.[72] The main thing now was to conceal the plan to seduce behind a utilitarian pretext. Perfuming linen was nothing more than a hygienic practice. Stationery emitted delicate scents; there was nothing to stop one thinking that they came from natural perfumes emanating from the writer.

Balzac was the self-appointed virtuoso painter of these vestibules and boudoirs where perfumes floated in the atmosphere without causing offense.[73] It was Madame de Sommervieux's skillful setting of the perfumed scene that overpowered Augustine and made her conscious of the gap that separated a draper's daughter of the rue Saint-Denis from a sophisticated aristocratic lady. The boudoir was depicted as the perfumed epicenter of the Balzacian universe; quite logically, since in the novelist's work, pleasant odor was most frequently associated with the words *flowers, women, Parisian, youth, loving, rich, clean,* and *unencumbered,* while stench was linked with the terms *confined, dirty, crowded, poor, old,* and *masses.*

Proximity to flowers, like proximity to birds, was innocent. "It is a taste natural to woman," the comtesse de Bradi confirmed; even prostitutes retained it when they fell from grace.[74] According to the Romantics, from Novalis to Nerval, the angelic, secret young girl, sensitive to the calls of the infinite, like the country flower opened a perfumed way to the other world of poetry. This proximity, this discreet harmony created the symbolic metamorphosis, kept alive the confusion. Well before Aurelia's silhouette was transformed into a flowering garden, Senancour wrote of the simple violet: "charms and swiftness of desires, with some small anxiety and some foreboding of the emptiness of things. Vague need to love; secret need to be loved. Delicacy in affections."[75] Rather than pluck it and absorb its perfume as sensual men did, Michelet asked the gardener-bridegroom to spare the poor flower, "leave it on its stem and raise it according to its nature . . . One needs to be grafted and a different sap introduced; it is still young and wild. Another is soft and sweet and entirely permeable; it only needs to be impregnated; there is nothing to be done to it but to convey life . . . its love-dust flies away in the wind; it must be well protected, held back, above all, fertilized."[76] The male's enjoyment was increased by his dreaming of fructifying the innocent corolla, in full bloom and consenting to his desire.

The flower's new authority was sanctioned by Ingenhousz's discovery of the process of photosynthesis. "The mephitic emanation from flowers and leaves is completely different," he wrote, "from the emanation that carries the perfume. The first is as much to be feared as the other is by nature innocent . . . The perfume of vegetables has nothing in common with their mephitic exhalation."[77] A fundamental differentiation; the plants that smelled strongest had ceased to be regarded as the most dangerous; flowers did not conceal a greater menace than leaves. To escape danger, it was enough to avoid sleeping near plants, to air places where they had been placed during the day, and to keep large leaves out of the house. Even better, by breathing over the flower in sunlight, woman could find a cure for her nervous frailty. "The plant, which has no nerves, is a sweet complement to her, a sedative, a cooling influence, a comparative innocence." Nevertheless, in his concern for morality, Michelet asked that "young maidens" avoid the compromising "confusion" of bouquets.[78]

The new theories prompted the strong return of flowers to favor. Their role in gardens had declined in the eighteenth century; in the gardens of the poor they only graced the borders of vegetable gardens at the most. Smell had played only an auxiliary role in the artificial landscape of English parks. Its function was solely to accentuate the impressions produced by the noble sensations of sight and hearing. The first nineteenth-century treatises on landscaped and picturesque gardens reveal an almost unchanged situation.[79] Innovation would come not from the untrammeled horizons of parks, but from greenhouses and the enclosed areas of bourgeois gardens.

The spread of greenhouses in the nineteenth century deserves attention from historians of private life. There were manifold varieties: winter gardens; hothouses sheltering exotic plants throughout the year; temperate greenhouses, heirs of the orangerie, where vegetation passed the winter protected from cold. Long reserved for the aristocracy and the very wealthy, greenhouses proliferated, particularly in England but also in central Europe and then in France.[80]

Experts required that they be contiguous to the house. People had to be able to reach them without exposing themselves to cold, still less to rain. It was a simple next step in the itinerary of perfume, designed by architects, for the greenhouse to open onto the "pleasure ground," called a kitchen-flower garden or a flower garden in France.[81] It extended the dwelling and bore witness to the expansion of the private sphere. Greenhouses provided a place to stroll in all weathers, which caused flowery arbors to be woven and benches placed in them. Greenhouses also became places for fortuitous meetings, rendezvous, adventures. They thwarted the surveillance that prevailed in domestic space. In summer, temperate greenhouses could also be used as resting nooks, reading rooms, dining rooms, even ballrooms.[82]

All things considered, greenhouses were not without their dangers, and it was important to control their "morbidness." If care were not taken, fermenting vegetation and rotting mold could create a dangerous swamp, a reservoir of miasmas on the threshold of the dwelling.[83] Ventilation was imperative.

In time, greenhouses became fashionable among the bourgeoisie. They were "an indispensable annexe to every garden of any size today," noted Baron Ernouf in 1868.[84] At that time the greenhouse-sittingroom described in the *Curée* was becoming popular in France.[85]

A subtle harmony was soon established inside the greenhouse between woman and scented flower, at least in central Europe. In extreme cases, the greenhouse invaded the home; plants and flowers scaled walls and staircases, penetrated boudoirs. Domestic space was identified with floral decoration; the atmosphere of the dwelling was redolent with vegetable perfume. The naturalist Bory de Saint Vincent described his astonishment upon entering Vienna in 1805:

> It was a new and entrancing thing for me to find the apartments of most of the elegant women there embellished with greenhouses, scented in winter by the most pleasing flowers. Among others, I remember with a sort of intoxication the boudoir of comtesse de C***, whose sofa was surrounded by jasmine climbing on a night-shade growing in earth, and all this on the first floor. The bedroom was reached through veritable bushes of African heather, hydrangeas, camellias, not at all common then, and other valuable shrubs planted in flowerbeds, studded with violets, crocuses of every color, hyacinths, and other flowers clustered in turf. Opposite was the bathroom, also built in a greenhouse, where papyrus and iris grew around the marble tub and water pipes. The double casements were likewise decorated with beautiful flowering plants.[86]

It was during the early nineteenth century that a garden aesthetics was formulated, intended for the upper—and then the lower—bourgeoisie. This development, somewhat overshadowed by the attention paid to foreign parks, is of major importance. The bourgeois garden was the landscape architects' solution for dealing with the problem of small areas. It would have been ridiculous to compete with nature by attempting to create a landscape. "Flower gardens and shrubberies (here a single scene) are the only suitable compositions for an area of not more than one arpent around a bourgeois house."[87] Unable, within such narrow limits, to subordinate the arrangements of the plan to the pleasures of sight and to draw inspiration fom the laws of perspective, the architects fell back on "smiling scenes," the only ones in landscaped gardens to give a considerable place to olfaction.

This work was accomplished at the very time that Gabriel Thouin was setting out to restore the abundance of flowers in parks by increasing the number of parterres. Some of his pupils, notably Bailly, applied the master's principles to the enclosed area of the bourgeois garden and attempted to codify the arrangement.

The landowner who was concerned about elegance had first to dissociate the flower garden from the kitchen garden; he had to abandon the practice—long-standing among the bourgeoisie—of growing flowers only in flowerbeds and borders.[88] The pleasure garden had to be enclosed by hedges, the kitchen garden by walls; it was important to eliminate confusion between the two separate areas of life. The area next to the house was similarly structured and differentiated at the same time that it was becoming an extension of domestic space. "It is like an additional apartment of the main building," Alexandre de Laborde noted of the flower garden as early as 1808.[89]

Time spent in the flower garden was organized in the same way and governed by the same requirements as inside the house. "Extreme" cleanliness had to prevail, as well as an "air of elegance and order." This garden "is like an apartment and is a gallery of well-arranged natural objects rather than an imitation of nature."[90] The rake played the same role as the broom.

There was one paradoxical aspect of this project: everything had to be done to extend possibilities for walking in this cramped plant-filled area. The garden had to combat the menace of a sedentary life, permit walking, promote good breathing. It became a labyrinth of winding alleys; these had been made fashionable by Thouin, anxious to break down, by a proliferation of round and curving flowerbeds, the old divisions into compartments and patchwork vegetation of French-style gardens.

A complex architecture of vegetation was employed to make the promenade more interesting and resting easier. Shelters of shady foliage, fresh and scented, and corridors of greenery encased a series of enclosed areas that guaranteed intimacy to the point where it became a threat to modesty. Like the greenhouses of the very rich, they became the only possible places for sudden seduction. The immense role of the garden alley in the private life of the bourgeoisie took shape at that time and then became more precise.[91] "It is there that a mouth, which is instantly tinged with the brightest rose, involuntarily makes the first avowal, the first fruit of happiness."[92]

Relatively little is known about these ephemeral green constructions of the July Monarchy, because their traces were obliterated by the belated proliferation of greenhouses and metal arbors designed for decorative plants, and then by the fashion for the artificial garden. Their existence suggests a whole line of research on the origins and forms of social appropriation of the vegetal.

A precise vocabulary was formulated that attempted to define these varied constructions of foliage.[93] The term *bower,* Boîtard stated, should be reserved for the "short covered alley" cut in the shape of an arch, impermeable to the sun's rays—and therefore to the eye. This miniature promenade would be shaded by honeysuckle, jasmine, or scented clematis, supported by a light wood frame. The arbor, most often circular, covered by a small metal dome, had a more substantial structure, but the same climbing shrubs covered it. The resting place most frequently consisted only of a simple stone bench, placed near a statue or bust, modest imitations, on a reduced scale, of the picturesque design of the English garden at the end of the eighteenth century. A clump of lilac or laburnum gave it shade. The most ambitious gardens also contained arbors, rooms made of vegetation, a ballroom, a dining room, even a theater built of greenery.

Conserving the limited amount of space, the walk provided as many pleasant smells as visual delights. In the absence of a landscaped horizon, it was up to flowers to gratify the sense of sight as well as the sense of smell. Before the fashion for artificial gardens gave rise to a proliferation of fountains and ornamental lakes, there were few pleasures for the ear in bourgeois gardens, apart from birdsong.[94] Sensory experience of the bower became all-important. The young girl learned to distinguish the discreet odors, the "mystery," and the language of "simple flowers" by walking in the garden, and not by breathing in the promiscuous smell of bouquets.[95]

What strikes the reader today is the delicacy of the perfumes emanating from the flowers and shrubs that the specialists recommended. The presence in the catalogs of the sweetest species corresponded logically to the near monopoly enjoyed by floral scents in the perfume trade. There were several favorites, some of which have since fallen into disuse. They included scented mignonette, the memory of which haunted Madame Lafarge confined in her Montpellier prison.[96] Its lack of beauty was evidence of the importance accorded to scent. Also popular were sweet peas, destined to become one of the flowers of the poor, basil, belle de nuit, and centaurea. However, the two undisputed queens of the bourgeois garden remained julienne and the common violet.[97]

Flowers also proliferated inside the home; they were no longer restricted to the woman's toilette. They decorated "floor boxes," "window boxes," and "massive vases of greenery."[98] There bloomed the roses, jasmine, lily of the valley, reseda, and violets that the arbiters of elegance recommended.[99] Exotic plants were judged too disturbing; it was not yet good form in France to transform the house into a museum of vegetation.[100]

During the Second Empire, fashion decreed that women decorate their clothes with flowers. "Natural flowers are used to decorate bodices; they are put on sleeves, often on skirts, not only in slashes and flounces, but in two or three rows in front."[101] Roses, wallflowers, lilies of the valley, jasmine, and forget-me-nots were artistically arranged in the hair and framed the faces of young, fashionably dressed ladies.[102] On the other hand, the code of good manners forbade the mature lady to use natural flowers. Harmony existed only between the young girl and flora. Those who had forfeited these youthful scents still had artificial flowers, but again, to be used with discretion.

The new craze stimulated the flower trade. The traditional flower market in Paris was no longer adequate; biweekly markets were organized in squares, then in boulevards. Along newly built passages "you need only shut your eyes in order to fancy yourself in a delicious flower garden," Mrs. Trollope observed in 1835, and she was little given to noticing anything in Paris that delicately caressed the senses.[103] These markets were more and more heavily patronized. As early as the beginning of the July Monarchy flower girls were legion on the bridges, along the quays, or on the pavements.[104] They created a new problem for moralists.

Flowerpots and bouquets spread to the masses, "down to the little working-girl who likes decorating her attic," noted Debay.[105] "They do not value the rarest," said Paul de Kock of the little milliners; "provided that they have stock or mignonette, they are satisfied; they stuff bunches of them into their carafes; all the same, they have to last the whole week and smell good."[106] The image of the beflowered little dressmaker was cheering. Scented with natural perfumes, the garret was the symbolic antithesis of the stinking hovel or licentious factory. The presence of flowers was evidence of a workplace that was consistent with the cheerful, clean, and hard-working young girl.[107] Even indoors, the innocence of flowers bore witness to virtue. Lit up and framed by curtains, the bouquet could, of course, be transformed into an inviting signal; clandestine prostitutes also knew the language of flowers.

Flora appeared less ambiguous in the countryside. Here, the models of embroidery from which young girls drew their inspiration in making their wedding trousseaux helped to spread the taste for innocent flowers; the lovingly embroidered corollas prepared the way for the furtive invasion of vegetable-garden borders.[108] The new pastoral theme that inspired the rural clergy—for which the curé of Ars was already the venerated model—centered mainly on young girls.[109] The children and servants of Mary made sure that her altars did not lack flowers; and if the parsonage garden was no longer adequate, nothing was easier than to plant more. This ensured that, when Corpus Christi came round, the young girls could fill their baskets with flowers in order to scatter them on the paths straddled by the dais.

Fantastic dialogues revealed the delicate affinities between the flower and the young girl in bourgeois gardens. Lilies, roses, and violets became secret confidantes, recipients, like the piano, of the impatient sighs of first emotions. Although the whiteness of the lily concealed disturbing scents, who could be offended by this innocent outlet for contained sensuality? While Gilliat fought the storm, the rugged reefs tearing his body, while the poor crowded into the stinking ship *Jacressade,* the little scented garden heard Déruchette's soul breathe and breed its chaste loves. The young girl watered her flowerbeds herself, Victor Hugo declared; her uncle "had raised her to be more flower than woman."[110] When she appeared in her garden at twilight, "she seemed the flowering soul of all these shadows."[111] During the spring, love had sharpened Gilliat's perspicacity and he was privileged to interpret this silent dialogue. "According to the flowers which he saw Déruchette pick and smell, he had discovered her tastes in perfume. Convolvulus was the odor she preferred, then pinks, then honeysuckle, then jasmine. Roses came only fifth. She looked at the lily, but did not smell it. From choice of perfumes, Gilliat created her in his thoughts. To each odor, he attached one of her perfect qualities."[112]

It is time that historians described the crowded *Jacressade* and the Promethean efforts of the man on his reef; by refusing to pay attention to Déruchette's breath, they are in danger of misinterpreting the dreams and desires of that fascinating, mad bourgeoisie that led the social game. The history of the mignonette, lily, and rose is just as informative as the history of coal. "A soft sweet perfume is an inexhaustible pleasure to her [La Fosseuse]; I have seen her take delight the whole day long in the scent breathed forth by some mignonette; and, after one of those rainy mornings that bring out all the soul of the flowers and give indescribable freshness and brightness to the day, she seems to overflow with gladness like the green world around her"; so Balzac fantasized on the mysterious harmony between the young girl and the breaths of nature.[113]

All that had ensured the dominance of perfumed flowers was eclipsed at the end of the Second Empire; the new aesthetics of parks, ordered by Napoleon III, revolutionized horticulture. "The taste for plants with beautiful foliage has been added recently to the taste for flowers," noted Edouard André in 1879.[114] Olfactory criteria no longer dictated the choice of species; the visual assumed primacy. Vegetation was selected for its majesty and its decorative effect when collected in a bunch. Highly colored plants were most appreciated.[115] Exotic species proliferated; the industrialization of horticulture, the creation of veritable "plant factories,"[116] ensured their triumph. The richest bourgeois conceived a passion for museums of vegetation, but the scents that filled these gigantic greenhouses no longer symbolized innocence. A new marriage took place between the fashionable lady and vegetation. Symbolist art provides abundant evidence of this trend, even before Mucha. Women surrounded themselves with lianas, poisonous in full bloom; they liked to live face to face with enormous corollas. They were no longer afraid of inhaling lilies, but flowers had ceased to be their confidantes.[117]

Short-Term Oscillations in Perfumery

The history of perfumery here could fill several volumes. I want only to emphasize certain major facts directly related to sensory history. From the beginning of Louis XVI's reign to Coty's compositions, the trend favored sweet floral scents. However—and even though there is no justification for talking about cycles—short-term oscillations in taste and fashion did occur; for example, every half-century, musk and ambergris unleashed short counteroffensives.

Under the Terror, choice of odors revealed political allegiance; perfume, given a new name, became a rallying sign. Smearing one's head with "pommade de Samson" was an affirmation of patriotic convictions. "You braved banishment and the guillotine," noted Claye, "by impregnating your ruffle and handkerchief with essence of lily or eau de la Reine."[118] After Thermidor, the penetrating perfume of the "muscadins" was evidence of reactionary sympathies.[119] The 1830 Revolution produced the same use of scent to signify commitment, ensuring the success of "Savon constitutionel" and "Savon des trois journées."[120]

The Directory, and even more the Consulate and the Empire, marked the return of strong perfumes with an animal base. The presence of aristocrats and the creation of the imperial nobility also helped to give perfumery a new impetus. The craze for things Greek and Roman resulted in the reappearance of unguents and perfumed baths. "Ancient oil, which was costlier than gold, anointed every head at that time. When Madame Tallien left a bath of strawberries and raspberries, she had herself gently massaged with sponges soaked in milk and perfumes."[121] All the witnesses agree that the court at the Tuileries was more intensely perfumed than the court of Louis XVI. A phial of the finest eau de cologne was poured over the emperor's head and shoulders every morning. Napoleon had a liking for vigorous massage. Josephine's taste for musk, ambergris, and civet is well known. The empress also had perfumes sent from Martinique. Sixty years later her boudoir at Malmaison still retained the odor of the musk that had saturated it.[122] The couple's correspondence shows the important role of bodily odors in their sexual relationship. This olfactory sensuality contravened the injunctions of the sanitary reformers; it was also distinct from the rose-water eroticism of people such as Restif de La Bretonne.

As we have seen, the Restoration also made a place for itself in the annals of odors. It inaugurated "the reign of old women," hostile to intoxicating perfumes.[123] Musk would have been out of place with the Touffedelys ladies at Valognes. Sensitive to sweet vegetable scents, the old-fashioned arbiters of elegance attempted to transmit their out-of-date tastes to their little girls. "Perfumes have gone out of fashion," noted Madame de Bradi in 1838. "They were unhealthy and ill became a woman, because they attracted attention."[124] The vague odor of powder "à la maréchale," which had impregnated her dead grandmother's apartments, awoke Louise de Chaulieu's moving memory of her childhood.[125]

196 ·

At this time too, the smell of tobacco became prevalent, as did the odor of camphor, which had an immense vogue.[126] Tobacco was recommended to the poor by doctors; François Vincent Raspail in particular extolled its preventive properties. It was chewed and smoked; invalids' beds were dusted with it; it was used for unguents, massage, and poultices.

In about 1840 the range of scents became more complex; the confrontation between flower and tobacco ended.[127] While male fashion matured, a new aesthetics of smell very tentatively emerged. Perhaps this should be seen as the result of neo-Lamarckism, which stressed the dangers of allowing functions to lie dormant.[128] Whatever the cause, thirteen years later perfumes triumphed at the court of Napoleon III, as they had under his uncle; but they were not exactly the same.[129] The series of data relating to the work force, production, and commercialization of products clearly show the rapid growth of the perfumery industry thereafter.[130] The introduction of chemical processes, the invention of the atomizer and, later, of the *hydrofère,* which made it possible to diffuse preparations in bathwater, all favored this growth.

Apart from eau de cologne, production was concentrated in Paris and London. The Exposition of 1868 was a triumph for the perfume industry of the two capitals.[131] Factories in Spain, Germany, Russia, and the United States no longer concocted anything but mediocre products. The signing of trade treaties had put an end to fraudulent imitations from the other side of the Rhine. Some businesses were dazzlingly prosperous. In 1858 the firm of Gellé already owned a factory at Neuilly as well as branches in St. Petersburg, Hamburg, and Brussels. The Parisian perfume industry used raw materials from all over the world and exported its products throughout the globe. Its main source of supply, however, remained the Grasse and Nice regions, along with England, which grew the most fragrant lavender. As early as 1850, trade with the Orient had been reversed; the Ottoman Empire now had a deficit. The most esteemed attar of roses now came from Paris.[132]

Preparations had been constantly improved since 1840. Haute parfumerie nurtured its future triumphs. The list of historic events that might explain this trend is long: the new rise of the fashion trades, the Bonapartes' return to the Tuileries, the triumphs of exoticism and cosmopolitanism, or, better still, Alexandre Dumas's untiring efforts to restore the taste for eighteenth-century perfumes; this last venture paralleled the revivalist efforts in the field of art of the Goncourts and those art collectors motivated by a new passion for the style of Louis XV.[133] The bourgeoisie could now uninhibitedly ape the aristocracy and pursue the accumulation of symbolic values. This is the underlying significance of the "imperial feast." Perfume profited from the temporary decline in diatribes against luxury and softness—and perhaps even more from the current quest for aesthetic syncretism. Baudelaire's theory of "correspondances" reflected something that was actually taking place in contemporary civilization. Whiffs of perfume scented the stage for English fairy plays. In Paris there was some idea of imitating this practice at the première of *L'Africaine.*[134]

In 1858 Worth created Parisian haute couture.[135] His salons were transformed into delicately perfumed hothouses and simultaneously reflected and gave new impetus to the stage set of the boudoir. But by that time there were already some great names in the perfume industry in Paris and London: Askinson, Lubin, Chardin, Violet, Legrand, Piesse, and, above all, Guerlain. Bouquets lost their simplicity. As early as 1860, Claye, blender for the firm of Violet, stated that perfumes required three or four years of research. Nevertheless, the new aesthetics were only tentative; they had trouble breaking away from the excessively rigid code established by the perfumers of the ancien régime.

The figure of the perfume-artist began to emerge. As early as 1855, Piesse suggested a scale of smells that aroused the mirth of contemporary chemists.[136] Here were perfumers daring to talk about harmony, perfect accords (sunflower/vanilla/orange blossom), dissonances (laurel/pink/thyme).[137] They appropriated the vocabulary of the masters of the Conservatoire, except that they offered no theoretical treatises, only practical measures. In fact the management of odors kept most of its secrets, and consequently most of its mystery. The sophistication of the flasks also revealed the new ambitions. The timelessness of crystal was allied with the evanescence of perfume, and Birotteau's erstwhile scientific strokes of genius would now provoke a smile.[138]

It was Huysmans who, in 1884, finally outlined the model of the modern perfume-artist. His hero Des Esseintes had a thorough knowledge of all its techniques.[139] His great composition emerged as if in an arranged sequence: it had a head and a base. Des Esseintes did not use a recipe; he let himself be guided by his poetic plan. He constructed a setting (the beflowered meadow), recreated an atmosphere ("light rain of human essences"), evoked sentiments ("laughter in a bead of sweat, joys disporting themselves in full sunlight"), inserted strident modernity ("the breath of the factories"). Twenty years later Coty created "Origan."

The vocabulary was refined in accordance with the new aesthetic claims. The wider variety of products and the search for comparisons called for a corresponding verbal effort of imagination. The immense range of brand names depicted a relatively simple poetic landscape in which a few major peaks stand out. The linguistics of bouquets confirmed the attraction for fleeting country odors (for example, "L'Heure fugitive"). Violets, roses, and lavender held sway over the vocabulary of scent. The Orient also kept its mirages. According to Rimmel, this was the result of the success of Niebuhr's *Description de l'Arabie* and numerous accounts of travels in Egypt.[140] Flaubert, encamped on the banks of the Nile, drew up an enthusiastic inventory of the perfumes of the desert.[141] The description of the bazaar at Istanbul strengthened the fascination with harems. On the other hand, the vocabulary of perfumery imposed a stereotyped image on oriental actuality. Only the name Constantinople, wrote Edmond and Jules de Goncourt of Anatole Basoche, "awoke in him dreams of poetry and perfumery where . . . all his ideas of Eau des Sultanes, pastilles of the seraglio, and the sun on Turkish backs mingled."[142]

However, the majority of names were connected with the aristocracy and the ruling families. Haute parfumerie thereby acknowledged its close links with European courts. Its spread was partly due to the immense popularity of royal couples in the middle years of the Third Republic. Political nostalgia sharpened the desire for luxury, while references to the princesses guaranteed the richness of the product. Adopting "Jockey Club" or "Bouquet de l'Impératrice," if not "Pommade de Triple Alliance," was to acquire an imaginary entry into the circle of prestigious lineages.

Over the decades, the aesthetics of the sense of smell became commonplace; the moderate price of perfumed soaps, the industrial manufacture of eau de cologne, the expansion of the network of drapers who distributed the products of perfumery enlarged the range of the clientele. Flasks began to adorn the shelves of doctors and minor provincial notables.[143] Even before toilet soap came into general use, the downward social mobility of eau de cologne was evidence that the poor man too had joined the battle against the putrid odor of his secretions.

12 *The Intoxicating Flask*

The Respiration of Time

THE SENSUALISM that prevailed almost universally in enlightened circles at the beginning of the nineteenth century encouraged glorification of sensory pleasure.[1] Contemporary works abound in references to the delights of smells, particularly in rural surroundings. Balzac is a good example of this awakened sensitivity, which attributed the origins of impulses and feelings to respiration of the perfumes of nature. Flowers, newmown hay (*Les Marana, El Verdugo*), the intertwined scents of the countryside (*Les Paysans, Mémoires de deux jeunes mariées*), and forests stimulated his heroes' sensual desire. "All these strong smells of fertility," declared Blondet, "fill your nostrils, bringing you all a thought, their soul perhaps. I thought at that time of a pink gown billowing through the winding alley."[2] Smelling the perfumes of spring, the young woman became sharply aware of her destiny.[3] The young Flaubert was overwhelmed by the less sharply distinguished odors of beaches and fields: the saline vapors of sea and seaweed, the scent of grass and the strong odor of dung increased his nostalgia for the Croisset of former times. His nostalgia was mingled with the Romantic fascination for the odors of excremental and cadaverous putridity.[4] Twenty years later, the Goncourts' hero Anatole Basoche, employed at the Jardin des Plantes, basked in "great animal bliss." His attitude showed the revival of the fascination with strong odors.[5]

Rather than prompting the "fleeting shock" that revealed the coexistence of the "I" and the world, the sense of smell was now geared to the harmonies between the moods of the individual and the odoriferous landscape. Closer attention was paid to transitory smells. The changing smells of hours, days, and seasons accompanied the "internal meteorology" that people such as Maine de Biran were trying to draw up in the fashion of Rousseau. Maine de Biran's wish to move away from Condillac's philosophy, his constant attempt at introspection, led him to transpose neo-Hippocratic self-monitoring into the realm of the experimental psychology that he was trying to create. "I have senses," he wrote in 1815, "that are extremely variable in their activity or their susceptibility to impressions. For example, there are days when the slightest odors move me; others (and these are most frequent) when I feel nothing."[6] His joy on the blessed days prompted him to record them. Thus, on May 13, 1815: "I am happy with the fragrant air that I am breathng"; and on July 13, 1861: "The air is fragrant."[7]

However, it was the aged Senancour who made the most penetrating comment on the harmony between seasonal variations in odors and movements of the soul. "The violet blooms in autumn too. It is the same odor and a different delight, or at least the violet arouses different feelings at that time; it provokes other ideas, it gives—less strongly perhaps—a more intimate, dreamier, less transitory satisfaction."[8]

Olfactory reminiscence now became commonplace. Maine de Biran meditated on the strange sensation that, according to him, tore off the veil between heart and thought, destroyed the distance separating past from present, and produced the melancholy of the "never more" through awareness of the unity of the "I." "The kind of memories that are attached to the sensations of the smell must be of the same nature as the sensations themselves, that is to say, purely emotional; there is an affinity between odors and the internal impressions that compose the feeling of coexistence, which is entirely peculiar to this sense. Odors, linked to such spontaneous, ineffable feelings as are experienced in youth, always awaken more or less the same feeling; you find yourself young again, in love, in a scented shrubbery. It is there that the heart plays out its game independently of thought; when the veil is lifted, we feel all that we have lost, and melancholy seizes our soul."[9]

The individual experience revealed by a few poets quickly acquired the status of scientific truth: the sense of smell was the sense of "tender memories," according to the *Dictionnaire des sciences médicales* in 1819.[10] In these accounts, the field of memory, like the gamut of permitted scents, was voluntarily restricted. In 1821 Dr. Hippolyte Cloquet waxed quite untypically lyrical: the spring odors of the forest, he confided, evoked "the image of a dear friend who is no more"; they were an invitation "to call to mind the glorious events of time past or to form plans for future happiness, unpoisoned by false resolves of ambition."[11] In a drier vein, Dr. Bérard wrote in 1840 in the *Dictionnaire de médecine* that the sense of smell "brings reminiscence and imagination into play."[12] No less magisterial were two lines in Balzac's *Louis Lambert:* "That sense in more direct contact with the cerebral system than the others must cause invisible agitation to the organs of thought by its changes."[13]

In the same way as Tennyson, Thomas Moore, and many others, George Sand indulged in the nostalgic pleasure of reminiscence. In an amazingly dense text, she linked memory of smells to the presence of the mother and to ontological feelings. "So, seeing the convolvulus in flower, she [the mother] says to me: 'Smell them, they smell of good honey; and do not forget them!' This is therefore the first revelation of the sense of smell that I remember; and by a link between memories and sensations *that everyone knows,* and cannot explain, I never smell convolvulus flowers without seeing the place in the Spanish mountains and the wayside where I first plucked them."[14]

To refer to the "deep magical charm with which the resurrected past intoxicates us in the present" was henceforth extremely banal.[15] Another example belongs to a different key, since it associates olfactory sensation with hearing. "I was still a child," wrote Alphonse Karr in 1870, "when my dear father composed a tune, much sung at that time, on a gloomy theme, the Barcelona plague in 1821. Well, when I hum those two verses . . . I get a definite smell of mignonette—in the same way that the smell of mignonette easily makes me think of the Barcelona plague and reminds me of the date."[16]

One final manifestation of this well-worn theme: it was the perfume of the woman in black that governed Rouletabille's police investigation in Gaston Leroux's novel.

Analysis deepened as the subject became more commonplace. Charles-Léonard Pfeiffer has detected the emergence of a "complex memory," in contrast to the simplicity of earlier recall. A short extract from *Madame Bovary* is enough to illustrate his point: "Emma, her eyes half-closed, took great breaths of the fresh wind that was blowing. They did not speak to one another, they were too lost in the invasion of their reverie. The tenderness of former times returned to their hearts, abundant and silent, like the river which flowed so softly that it carried the perfume of syringa and projected into their memories shadows more excessive and more melancholy than those of the immobile willows which lengthened on the grass."[17]

Even more subtle was the pleasure of Fromentin's Dominique, for whom reminiscence of the smells of sensual delights acted as a substitute for what the presence of his beloved Madeleine could not supply. "The tiniest individuality of her dress or person—an exotic perfume she loved, which I should have known with my eyes shut— everything, even the colours she had affected for some time . . . everything came to life with surprising vitality, causing me, however, an emotion other than her presence would have done, a regret, almost, to be embraced with pleasure, for all the lovable things no longer there."[18] For the rejected lover confined to the countryside, winter was the favorite season for this hedonistic remembrance of sounds, visions, and odors.

The timelessness of perfume, a theme dear to Baudelaire, endowed the sense of smell with an overwhelming evocative power. What would remain of the man and his loves to those whom he had cherished? A perfume imprisoned in a flask, an odor lodged in the back of a cupboard or tomb.[19] Breathing certain scents could resuscitate a society, a former civilization. A respectable old lady was disturbed to find in the Saint-Cloud of Louis XVIII the excremental odors that prevailed in the Versailles of Louis XVI. She confided to Viollet-le-Duc that this mark of aristocratic indifference to the stench of excrement awoke nostalgia for her lost youth and the vanished days of the ancien régime.[20] Théophile Gautier, obsessed with retrospection, counted on help from the rocklike reliability of ancient odors to "transpose his soul."[21] Incense defied time by virtue of its resistance to decay; sensitive worshipers saw the sacrifices of the past rise up before their eyes through its sacred effluvia. "The acrid odor of time" was overwhelming.[22] Perfumery manuals came to resemble history books. Claye, who wrote such a manual, confessed that he was very conscious of the confusion between the passion for odors and the vertiginous depths opened to the historical imagination. Des Esseintes undertook to resuscitate the past by scientifically creating the smells of its environment; history was destroyed by the recreated odor. The act of breathing attested to the unity of time as well as to that of the "I."

The Censer in the Alcove

Since scientists laid it down that every individual possessed a specific odor, smelling oneself, examining the changes in the smells of one's own body, was already to become aware of the nature of one's being. By breathing the effluvia of her puberty Pauline Quenu experienced her destiny as woman.[23] The "grown girl's" narcissism found its solitary delights in olfactory as well as tactile self-regard.

Phrenology exaggerated these beliefs; in phrenological terms, "odor is the manifestation of beings, like line, color, and resonance."[24] The influence exerted by the discipline, particularly on Balzac, is well known. Birotteau, the virtuous man, chose to become a perfumer; Roguin, who was evil, became "foul-smelling."

Medicine and phrenology, both of which urged personal hygiene, went on to guide erotic behavior. If it was true that odor revealed so much about individuals, breathing the smells of other people assumed incalculable significance. The sense of smell was both the sense of social repulsion and the sense of affinities. The delicacy of the perfumed messages, the whiteness of the skin, the airy toilette were invitations to breathe in the smell of the woman.[25] Remembering the smell of the beloved's body kept passion alive and nourished regrets. This subtle observation was practiced only by the bourgeois. The effluvia of the fragrant mistress belonged to the repertoire of sentimental education.[26]

In this respect Balzac's work reflected both the medical beliefs and the code of elegance of his day. Fascinated by the seduction exercised by the messages of smells, the novelist made *Le Lys dans la vallée* a symphony of "perfume appeal."[27] "She took a few light steps as if to flutter her white toilette . . . 'Oh, my lily!' I said to her, 'still, perfect, and erect on its stem, still white, proud, perfumed, solitary!'"[28] Félix de Vandenesse seems to have been inspired by Cadet de Vaux. In Balzac's narrative, the natural odors of the woman's body seduce through their floral delicacy. Quantitative analysis shows that discussion of smells focused on hair and, secondarily, on the exposed parts of the body. The new code of hygiene ordained that they be kept clean, and modesty did not forbid mention of their odor, since it asserted itself in social relationships (neck, décolleté, bust, arms, hands, face). There were very few allusions to the perfumed odor of hips and waist.

With Baudelaire disappeared that poetic harmony between the woman and the flowers of the field that in the erotic realm had taken the place of the insistent presence of the scented shrubbery, the customary backdrop to love scenes of earlier times. The scented profile of the woman was transformed; she was no longer delineated beneath filmy gauze. The perfume of bare flesh, intensified by the warmth and moistness of the bed, replaced the veiled scents of the modest body as a sexual stimulus. The visual metaphor died out. The woman stopped being a lily; she became a perfumed sachet, a bouquet of odors that emanated from the "odorous wood" of her unbound hair, skin, breath, and blood.[29] The woman's perfume set the seal on the erotic intimacy of the chamber and the bed. She was the "censer" in the alcove,[30] exhaling a whole cluster of scents—the negative equivalents of which were stale tobacco and, even more, the musty odor of rooms, which attested to her absence. The emanations of the flesh gave life to the home and made it the theater of ceaselessly clashing smells. The atmosphere of the alcove generated desire and unleashed the storms of passion.

Baudelaire's poetry reflected both the movement of fashion toward heavy scents and a new significance attaching to sexual venality. The attractions of moist flesh, the poet's taste for animal perfumes, and, perhaps even more, his repulsion at lack of intimate hygiene[31] transposed the effluvia and scrupulous toilette of the brothel into the domestic sphere.[32] The judges never forgave this transfer of the erotic scene.[33]

Paradoxically, Zola had a weak sense of smell.[34] When he was subjected to Jacques Passy's olfactometer, the novelist performed poorly.[35] Léopold Bernard, though not aware of this, has already discerned the emphasis on smells in Zola's novels as a technique of naturalist writing.[36] Alain Denizet drew more precise conclusions from a detailed analysis of the *Rougon-Macquart* cycle.[37] Zola transposed into novels—very belatedly— the obsession with smells that had haunted medicine before Pasteur. His descriptions of the odors of public and private places, of the dwellings of both rich and poor, reflected the sort of obsessions found in the writings on sanitary reform around 1835 after the great cholera morbus epidemic. His extremely precise descriptions of the smells of individuals were similarly inspired by out-of-date beliefs. His systematically constructed correspondence among places, feelings, and loves seems to reflect the outcome of the patient work by sanitary reformers, architects, and artists that resulted in the fragmentation of the framework ordering the smells of intimacy. Perfumes orchestrated both the tender revels of Cadine and Marjolin and the progress of Renée Saccard's passion. The atmosphere of the rooms where she carried on her love affair with Maxime governed the shades of her feelings and pleasures; it was among the disturbed odors of the greenhouse that she experienced greatest sensual intoxication.

But there was more to Zola's works than an addiction to the past. By revealing to the hero his desires and his deeper nature, the messages of smell launched or restrained the action. Léopold Bernard has already noted concerning the characters in *Rougon-Macquart* that an impression received by the sense of smell was most often "the first principle and the last reason, conscious or not, for their conduct."[38] Baudelaire was not forgiven for transposing the permissive, heavy atmosphere of the brothel into the domestic context; no more was Zola pardoned for the dramatic role he gave to odors. By placing sight and hearing, the intellectual and aesthetic senses, on the same level as smell and touch, the senses of vegetable and animal life, he probably threw down his most scandalous challenge.

In Zola's world, sensory methods of seduction varied with social class. Touch prevailed among the masses. In both town and country, contact with the body, its contours clearly apparent, opened the floodgates of desire; the male seized his conquest. Among the bourgeoisie, olfaction governed the development of impulses and feelings. The obstacles placed in the way of sight meant that the charms of the body concealed from even the most furtive tactile contact, had to be guessed at.[39] The smell of the opposite sex unrestrainedly solicited the imagination, revealed affinities, made the blood seethe. With the oblique collaboration of the surrounding atmosphere, they found alliances.

A New Conduct of the Rhythms of Desire

The primacy of touch among the masses meant that the assault was brief; the subtlety of the messages of smell was in accordance with the delayed pace of bourgeois seduction. The evanescence of perfumes encouraged the relishing of anticipated enjoyment, intoxication; it symbolized the discontinuous nature of the dialogue of love. Patiently breathing in the breath of the loved one presaged the delicacy of caresses to come.[40] Like voyeurism, some olfactory behavior patterns permitted a new conduct of the rhythms of desire. Sniffing perfumed objects was an even better way of ensuring the imaginary presence of the mistress than looking at a photograph. This seizing the breath of the other person from a distance corresponded to Flaubert's limited and discontinuous love for Louise Colet. It was the quest for mysterious contact that enabled Frédéric Moreau to live in Madame Arnoux's atmosphere and Léon to experience Emma's reverberations.[41] In the course of the correspondence, scented letters, slippers, handkerchiefs, mittens, and hair built up a copious collection of smells. A practice of associating sight with the sense of smell emerged and rapidly turned into a ritual. It is described in minute detail in Flaubert's *Selected Letters*. A few extracts from August and September 1846 will suffice.

August 6: "I look at your slippers, your handkerchief, your hair, your portrait. I reread your letters and breathe their musky perfume."

August 9: "I'll take another look at your slippers again . . . I think I love them as much as I do you . . . I breathe their perfume, they smell of verbena—and of you in a way that makes my heart swell."

August 11: "In daydream I live in the folds of your dress, in the fine curls of your hair. I have some of those here: how good they smell! If you knew how I think of your sweet voice—of your shoulders and their fragrance that I love."

August 13: "Your mitten is here. It smells sweet, making me feel that I am still breathing the perfume of your shoulder and the sweet warmth of your bare arms."

August 15: "Tell me if you use verbena; do you not put it on your handkerchiefs? Put some on your slip. But no—do not use perfume, the best perfume is yourself, your own fragrance."

August 27–28: "Thank you for the little orange blossom. Your whole letter smells pleasantly of it."

August 31: "Thank you again for the little orange blossom. Your letters are perfumed with them." And, in the form of a bouquet, on September 20: "A thousand kisses . . . on those long curl papers; I sometimes breathe a little of their odor in the small slipper with the blue slashes, because it is there that I have packed away the lock of hair; the mitten is in the other one, next to the medal and beside the letters."[42]

References to smells proliferated in the course of correspondence when the rapture increased or when the amorous invitation became more urgent. Nearly half a century later, this type of erotic behavior would be described as fetishistic or neurotic; by then it had become more difficult to acknowledge such conduct outside the psychiatric sphere. Zola took it as the central theme for *La Joie de vivre,* a novel about the primacy of olfactory desire in a neurotic, degenerate individual. On several occasions the perfume of heliotrope emanating from Louise gives the action new impetus. It is sensitivity of smell that explains the good Veronique's discernment when she informs Pauline about the young couple's love; it is the odor of the glove discarded by Louise that ravages the unhappy Lazare for weeks on end. "The glove made of Saxon skin had retained a strong odor, the odor of a particular kind of deer that the young girl's favorite perfume, heliotrope, sweetens with a touch of vanilla; and, very sensitive to scents, violently disturbed by this mixture of *flower and flesh,* he had remained distracted, the glove to his mouth, drinking in the voluptuousness of his memories . . . When he was alone, he picked up the glove again, breathed it, kissed it, believed that he was holding her in his arms again." Fond of "wallowing in the burning memory of the other," Lazare wears himself out with these "veritable debaucheries."[43]

The sense of smell is implicated here in the sensitive issue of autoeroticism a short time before the aphrodisiac property of the smell of leather stimulated a copious sexological discourse.[44] But Edmond de Goncourt had already given Zola's Lazare a sister.[45] Chérie, still a young girl, has a mania for perfumes. She has even procured the forbidden enjoyment of a grain of musk, which she breathes secretly in bed. She makes herself so drunk on it that she induces an orgasm. Chérie's mother has gone mad; she does not want a husband. Ill informed, she thinks she can conceive without a man. The strangest of Edmond de Goncourt's heroines meets the gloomy fate doctors decreed as inevitable for women who masturbated; she will die a virgin without having known any pleasure except this curious substitute.

When the novel appeared, psychiatrists had already been striving to codify olfactory fetishism for several decades. In 1857, Tardieu used Latin to describe the sordid practices of "sniffers" who derived their pleasure from sniffing women defecating.[46] Ten years later, the policeman Macé described the astonishing behavior of these "sniffers," "lovers of ringlets," or handkerchief thieves, who swooped down on customers in big stores in order to smell the odor of their perfumed napes for a few seconds.[47] Féré analyzed the role of odors in unleashing desire.[48] Binet studied fetishism and examined Restif's behavior.[49] The sexologists Fliess, Hagen in particular, and, later, Havelock Ellis studied the immense role of olfaction in sexuality. But the person who introduced this new period was Huysmans. His hero Des Esseintes had heralded the perfume-artist's entrance into the world of art. Now, in *A rebours,* one of Des Esseintes' former mistresses displayed all the symptoms of the most unbridled smell fetishism. She was "an ill-balanced, nerve-ridden woman, who loved to have her nipples macerated in scents, but who only really experienced a genuine and overmastering ecstasy when her head was tickled with a comb and she could, in the act of being caressed by a lover, breathe the smell of chimney soot, of wet from a house building in rainy weather, or of dust of a summer storm."[50]

Above all, Huysman's book challenged the hierarchy of smells that had been accepted dogma for over a century. Des Esseintes was a collector of flowers that looked artificial; he denied the attraction of natural scents. Fascinated by Pantin's countryside, perfumed by industry, he suggested a new relationship to nature and glorified the odors of modernity.[51]

At the time Huysmans was writing, the history of olfaction was undergoing major change; it was burdened with the predominant anxieties of the time. Criminal heredity and the threat of animal regression would soon be implicit in Gaston Leroux's work. Rouletabille owed to his bandit father the animal sense of smell that he used to such good effect; his mother's disturbing perfume and great beauty did not prevent him from sinking to the ground and sniffing the earth on all fours to solve the criminal puzzles submitted to him.[52]

Post-Darwinian anthropology emphasized the specific odor of different races or ethnic groups. Jean Lorrain found the odor of Negroes, crowded onto the Champ de Mars, offensive.[53] Bérillon considered that this odor kept racial hatred alive in the United States and also formed the basis of segregation.[54] Before the patriotic outburst against the odor of the "Boche," Dr. Cabanès described the stale odor of the English, which he claimed so permeated their bedrooms that it remained there for several years. Some people, he stated, attributed this odor to proximity to algae and seaweed, others to the leather of their luggage. The Japanese scientist Buntaro Adachi denounced the stench of Westerners, and it was left to Dr. Bérillon to conclude, on the subject of racial antagonisms: "Nothing can prevail against aversions to smells." This led him to place an extraordinary value on olfaction, which he regarded as the favored instrument for the preservation of the race, "the formation of a definite family and the solidarity of the family environment being under the undeniable domination of the affinity and sympathy of smell."[55]

One less delicate subject remains: the evocation of the perfumed woman by, say, Houbigant's "Quelques Fleurs" (1912) or Guerlain's "L'Heure bleue" (1913). Thanks to haute parfumerie, a new code of olfactory elegance was taking shape, while the whiteness of woman and her floral symbols were reworked in a new scenario.

*"Laughter in
a Bead of Sweat"*

BEFORE THE TRIUMPH of Pasteur's theories, deodorization tactics were aimed essentially at public space, the common areas of insalubrious premises, and the dwellings of the rich.[1] The great majority of the population had no wish to know of the work in progress. The masses had hardly any exposure to the new discipline except through hospitals, prisons, or barracks. The dissemination of the codes of hygiene scarcely entered the educational realm before the 1860s.[2] Later, conscription laws, the normalization of the conditions of school life, and the persuasiveness of the Pasteurian credo allowed the value symbols and behavior patterns defined previously to spread slowly. Thus there is nothing surprising about the persistence of traditional behavior, including the resistance to strategies of deodorization efforts. The disappointments experienced by municipal officials and sanitary reformers in their battle against dung, filth, and vitiated air bore witness to loyalty to an ancien régime of sensory values.

*The Battle
against Excrement*

The impossibility of installing sewerage was evidence of the keen resistance in France to the policy of distancing man from human excrement, dung, and rubbish. There is a multitude of explanations available. First, it was based on Western scientists' ancient and abiding belief in the therapeutic value of excrement; in Madrid before Aranda's ministry, fecal matter was thrown into the streets. According to Chauvet, doctors claimed that this stench, which extended over more than four miles, preserved public health. Without the odors of sewage, he added, "we would soon have the plague."[3] At all events, this was the opinion of some experts; several even suggested spreading excrement in the streets of towns ravaged by epidemic. Fourcroy had doubts about these alleged properties of refuse but did not yet dare to deny them openly.[4]

On occasion, such beliefs guided therapeutic practice. During the reign of Charles II the authorities in London ordered that all the cesspools in the city be opened in order to conquer the plague by means of unpleasant odor. This Hippocratic atavism is described without a trace of derision in the 1787 *Encyclopédie méthodique*.[5] Half a century later, Parent-Duchâtelet was still praising the therapeutic qualities of excrement; according to him, they accounted for the good health enjoyed by gut dressers and sewermen.[6] Three consumptive women had been cured by taking part in the processing of the material. Parent-Duchâtelet found them "notable for their complexions and plumpness." "I knew," he added, "that several invalids who had been brave enough to immerse either a limb or their whole body in the vilest tanks had found a cure there, either for pains in the legs or rheumatisms or other infirmities that had withstood all other methods."[7] The water that flowed out of the Montfaucon basins was given as a cure to neighborhood horses.[8] Liger was still mindful in 1875 that cholera never raged in the vicinity of the Bondy sewage dumps.[9] Some practitioners considered excremental stenches troublesome but not unhealthy.

It is true that these theories were held by only a minority of doctors; most were convinced of the dangers of putrid infection. But the minority's theories strengthened the masses' belief in the beneficial property of refuse. Bailly noted in 1789 that butchers attributed their general good health to inhaling the odors of the blood, fat, and entrails of the animals they slaughtered.[10] In 1832 workers at the Montfaucon refuse dump remained convinced that excremental emanations were beneficial to their health.[11] Twenty years later, in an investigation of cesspool-clearers, Bricheteau stated that they did not regard the excremental odors they were exposed to as unhealthy.[12] He also emphasized how easily these workers found wives and mistresses.

But excrement found other allies. Manure merchants, agricul-turalists, and chemists cried out that deodorizing it was equivalent to impoverishing it. This drop in quality discouraged purchasers and resulted in a lower value for the product.[13] For that reason, the measures decreed by the municipality of Lille for disinfecting cess-pools encountered opposition from those in the trade in 1858.[14]

Bourgeois deodorization presupposed wealth, or at least comfort; it attested lack of involvement in manual labor. The poor man, the dung-man, permeated by odors, justified his rejection of deodori-zation in terms of his wish to survive. The peasants were determined to keep the indispensable manure at their doors.[15] In Paris, ragpickers opposed the municipal measures.[16] At the beginning of the July Monarchy they unleashed veritable riots against attempts by the pre-fecture of police to accelerate the removal of filth. From April 1 to April 15, 1832, ragpickers obstructed the movement of the filth contractors' vehicles and burned the new dungcarts.[17] The rioters were helped by the crowd; they too were worried about the disin-fection measures. The extensive use of chlorine in water multiplied rumors; some people saw it as evidence that the elite were bent on mass homicide.

A better understanding of this loyalty to filth depends in part on the role played by excrement in infant psychology and the importance of anality in the development of the psyche. It was by odor that the babe in arms experienced the mother's presence, before even seeing her. It was through the difference between hearing and smelling that he took the measure of space. Finally, it was through the effluvia they emitted that the tiny child distinguished between men and women. The odor of the baby's feces was a summons to the mother; in his dealings with her, the baby "produces something to smell from below and submits to something to smell from above": breast or bottle.[18] The spread at the beginning of the twentieth century of English-style toilet training, which forbade the child to walk about with a bare bottom or to defecate or urinate when he felt the need, and be cleaned up quickly by his attendant, completed the gradual rise of the discipline of defecation.

The role of odors in the awakening of sensuality should probably also be mentioned. Yvonne Verdier pursued an interesting hypothesis when, in discussing the foresters of the Châtillonnais, she stressed "the role of excremental odors in the formation of male erotic sen-sibility."[19] All the information we have about the sexuality of the masses in the past century comes from refined members of the bourgeoisie, who were not in a position to understand the impulses of those who did not share their repulsions. The attachment of the great mass of the population to strong, foul odors—despite the injunctions of the privileged classes—might provide a mode of access to the history of social psychology.

Most contemporary discourse associated scatological behavior with instinct, that is, with childhood and with the masses; it contrasted it with the behavior of the educated, mature bourgeoisie, who had been able to assimilate the somatic discipline necessary to remove excrement from the realm of sight and smell. While aristocrats continued for some time to show relative indifference to the injunctions of this chapter of the code of good conduct, the masses were set on publicly displaying their allegiance to filth. They proclaimed their prejudice in favor of degradation as opposed to sublimation, which was the aim of the bourgeoisie.[20] Some scatological practices—such as throwing excrement and waste, which was a feature of Shrovetide battles, or farting audibly, sometimes with accompanying gestures— revealed the masses' desire to let off steam. It was the antithesis of the process of accumulation that occurred in the cesspool. This prodigality, it was said, was evidence of the rejection of fecal discipline and, more generally, of "the obliteration of the Dionysiac function of the body";[21] unless these excesses were only temporary outlets, allowed as a concession by the disciplinary process in order to secure its own efficacy.

Even more evident was the resistance to the movement for the deodorization of language, which had been launched at the beginning of the seventeenth century. Invective punctuates this foul-mouthed, little-studied literature, which reached its peak at exactly the time when deodorization tactics were being implemented. The terrible dangers of decay, denounced by the experts, show themselves here, though in different terms, as a veritable obsession. "Filth, decomposition, the odor of rot, carnage, mess, the rotten, waste, rubbish, the dustbin, the cesspool, the sewer were evoked turn and turn about in an endless variety of images. Human excrement was only one form of filth, only one particular kind of waste."[22]

Perhaps the masses' fascination with decay was only their version of the obsession with putrefaction that overwhelmed the ruling classes. But there is another interpretation. "The king's pure language," noted Dominique Laporte, included a "low language" that would be "the place for verbal filth."[23] The scatology of the Shrovetide Carnival, the derision directed at hygienic consciousness, and the streams of abuse might be interpreted as acceptance of an allotted role in society. The masses, aware of the difference in thresholds of tolerance to smell, assumed the existence of this division and set out to align themselves openly against the deodorization practices. Throwing filth or its verbal equivalent became an acknowledgment of a position as much as a rejection of discipline. By throwing his excrement, the "little man" was doing nothing more than throwing a challenge to the people who avoided contact with him, in the same way that they avoided contact with excrement; he strengthened his own excremental status by his actions and words.

Two Conceptions of Air

The rejection of ventilation was part of this resistance to the enterprise of deodorization. In the countryside, in an environment in which the idea of the individual was not yet pertinent, inhaling the atmosphere of the family group, animals and people intermingled, was considered reassuring, like the warmth of the collective bed in winter. The proximity of animals was still accepted, even sought, when wakes were held. This attitude was strengthened by doctors' long-standing attribution of beneficial qualities to the air of stables inhabited by young animals. This alleged therapeutic quality perpetuated a lively polemic at the beginning of the nineteenth century. Most sanitary reformers disputed it, as the report of the Parisian experts clearly showed. But there were still important figures among advocates of the theory. Hippolyte Cloquet, the great patron of osphresiology, rallied to the cause while still showing his attachment to the precepts of hygiene: the air of stables was salutary, he wrote, as long as the animals were kept clean.[24] Medical archives yield abundant proof that this vitalist aerotherapy continued to be prescribed well into the nineteenth century. Many consumptives were sent off to breathe animal exhalations.

These factors explain the flagrant rejection of fresh air by the masses, particularly by the old, cold people who loved cozy nooks. "The masses are very fond of closed curtains and windows," Fodéré deplored.[25] Senile schoolmasters, anxious to breathe their pupils' odor, refused to open classroom windows.[26] "Our poor workers, being accustomed to enclosed dwellings, do not at all like the air to be changed when they leave them to go to hospitals or workhouses," Howard observed.[27] Old people around Les Halles refused to air their bedchambers, observed Dr. Legras in 1818.[28] Dr. Gregory, a Scottish sanitary reformer, settled the question authoritatively: "In visiting the poor [he] used often to begin his prescription by breaking a pane or two of the windows with his walking stick."[29]

A similar attitude was displayed in hospitals. Doctors at the Hôtel-Dieu at Lyons "preserved an invincible prejudice against the free circulation of air."[30] In some London hospitals and in the hospitals of Pamplona, people refused to wash rooms and open windows.[31]

The resistance to ventilation formed part of a wider resistance to new authorities. Hospital administrators' efforts at discipline proved futile. Recent historical studies of hospitals and prisons have stressed the dichotomy between the severity of the regulations and the anarchy of actual behavior patterns. The strength of the countervailing pressures exerted within those establishments, where the victory of the new disciplines was confidently awaited, become more and more evident. Inside the civic almshouses of Lyons during the Restoration, old men smoked and gambled, children ran wild, and the infirmaries looked like taverns.[32] Because of high demand, several inmates were crowded into the same bed and the space between beds reduced. Goaded by doctors who were scandalized by the insalubriousness of the place, the authorities strengthened their efforts to disinfect the hospital during the July Monarchy. The administration struggled vigorously to impose order and hygiene; it installed clocks and urinal cubicles and forbade unscheduled visits. After a delay of fifty years it strove, though again with no great success, to achieve the control of air and water flows that the reformers of the ancien régime had called for.

At the other end of the social spectrum, the bourgeoisie resisted ventilation inside their dwellings, but for different reasons. The withdrawal to the conjugal hearth, the rise of narcissism, and the phobia about importunate contacts and indiscreet odors, all aspects of a new way of life, ran counter to the demands for fresh air. We have already seen, in relation to the bedroom, what a subtle equilibrium sanitary reformers strove to maintain between, on the one hand, the requirements of health, which urged that all windows be thrown open and bed alcoves abolished, and, on the other, the joys of intimacy, which encouraged proliferation of curtains, hangings, and draperies. The norms of temporary ventilation to which the housekeeper had to

conform made it possible to preserve the salubriousness of the pad-
ded atmosphere of the *fin de siècle* dwelling and still prevent the
infiltration of miasmas from the street. Thanks to invisible, inaudible
servants, Des Esseintes could nurse his neuroses and enjoy his col-
lections without risking asphyxiation.

The Virtues of Filth   The pace at which the code of good manners
spread to the different social classes is ex-
tremely surprising. For example, the fastidious Dr. Freud was disturbed at the idea of
mounting the stairs between his bedroom and
his study minus his butterfly collar, but he had no scruples about
spitting on his bourgeois patients' stair carpets.[33] Théophile de Bordeu, anxious to see hygiene diminish the *aura seminalis* of his city-
dwelling clientele, attributed the great fertility of the poor to the
aphrodisiac effect of strong bodily odors.[34] Doctors at the hospital
in Amsterdam, noted Howard, regarded clean linen as unhealthy.[35]
Moreover, the reservations displayed by sanitary reformers toward
baths are well known. That the majority of the population long
remained convinced of the virtues of filth, however malodorous,
should not, therefore, occasion any surprise.

Françoise Loux and Philippe Richard have analyzed a corpus of
several thousand proverbs and shown clearly that the resistance to
bourgeois norms, displayed in peasant society, hid other norms, no
less precise but much harder to detect.[36] In matters of bodily hygiene,
sanitary preoccupations based on out-of-date medical beliefs and the
wish to preserve a primitive comfort triumphed over respect for
convention.[37] The physiological necessity of excretion governed behavior patterns. The proverbs advised against retaining either belch
or fart. They recalled the contagiousness of the desire to urinate.
They spun a network of prohibitions around baths, viewed merely
as a means of refreshing the body and not as a hygienic practice.
They recorded the erogenous function of bodily odors, to which they
only rarely attached a stigma. Proverbial discourse on cleanliness
emerged as an ethical discourse that extolled the evacuation of unhealthy humors or the odor of shirts and acknowledged that urinat-
ing, like drinking, gave new impetus to male social life. Even more
surprising, several proverbs stressed, either literally or metaphori-
cally, the connection between money and excrement later detected

by psychoanalysis. The coherence of this normative system, stressed earlier by Luc Boltanski, helps to explain the delay in the metamorphosis envisaged in the school and regiment.[38]

Dirt could satisfy the canons of beauty, which demanded paleness. It alone was capable of protecting the peasant woman, exposed to the heat of the sun, from sunburn. "Beautiful complexions are formed under dirt."[39] "The dirtier children are, the healthier they are."[40] There is no need to repeat what has been written so often about the prohibitions surrounding menstruation hygiene and, more generally, woman's intimate toilette. Less well studied but equally influential were the obstructions to the progress of hygienic reform played by certain forms of spirituality. Benoît Labre, like the Fathers of the Desert, who were fascinated by excrement, fed on vermin from his own body; according to Philippe Ariès, he believed "in the virtues of filth."[41] Fifty years later, his disciple Jean-Marie Vianney, the curé of Ars, shared his indifference. The curé's excessive behavior makes it easier to understand this attitude. What was the good of caring for the body that the holy priest flagellated, tortured, and called his corpse? Obsessed by the example of the great ascetics of the past, inspired by the *Légende dorée,* the curé of Ars refused to allow anyone at all attend to his home. He gave his clothes to the poor, neglected to change his cassock. All he was interested in was the "household of the good Lord." Humility drove him to seek out stenches, harbingers of the fate in store for the skin he was in such a hurry to leave. Jean-Marie Vianney took part in clearing the cesspool at his school. He followed the barrel-cart on its way to empty the material into the sewer.[42] Information supplied by his followers attests to his lack of dental hygiene and his bad breath. The attitude of the curé of Ars helps us understand the reserve displayed toward bodily hygiene by numerous parochial schools.[43]

The Libertinage of the Nose

The favorite object of both abuse and anger, adopted by everyone who claimed to reject the bourgeois norm, was the new repulsions. The provocative emphasis of this discourse and the way in which it focused on excrement and putrid odors attested to the high stakes involved.

The challenge to good manners issued by the young Flaubert was more virulent than his later denunciation of received ideas. He called for active overthrow of the code, notably at the level of smell; "Let diarrhea drip into your boots, piss out of the window, shout out 'shit,' defecate in full view, fart hard, blow your cigar smoke into people's faces . . . belch in people's faces," he advised his friend Ernest Chevalier on March 15, 1842.[44] Feces played a star role in his Rabelaisian schoolboy verbal revolt; he even referred to it in the polite formulas with which he ended his correspondence, although its potentially scandalous effect was diminished by the involvement of his male correspondents in the same game. Flaubert, so repulsed by the stink of the proletariat but very conscious of the role of anality in the emergence of narcissism, promoted excrement as a symbol of the "I."[45] In maturity, the author of *L'Education sentimentale* appreciated the vulgarity of prostitutes, their indifference to "coarse words," and their refusal to pass over physiological needs in silence.[46] He inspires speculation about the roots of the attraction that led members of the bourgeoisie at that time to line up for obscene and salacious visits.

The historian Michelet was obsessed with organic time, with the history of flesh that bloomed and rotted. He did not recoil in terror from putrefaction and the products of excretion. He was watching for the moment when the excreta, immediately after leaving the body, barely caused repugnance; he sought in them the traces of the outflow of life. Thus it is not surprising that the greatest of historians extolled his young Athenais's periods or filled his lungs with the musky odor of latrines in order to renew his inspiration.[47]

With Vallès, revolt broke out. One has only to read *L'Enfant* to realize that his "libertinage of the nose" is neither pure provocation nor fascination with death.[48] In this novel Vallès takes pleasure in displaying his olfactory sensuality. Jacques Vingtras's behavior puts him far beyond the bounds of good form: "I open my eyes enormously wide, I flare my nostrils and I prick up my ears"; even better: "I opened my nostrils wide . . ."[49] According to his personal criteria and without reference to good manners, the youth creates his own hierarchy of smells, one far removed from the hierarchy defined by fashion. His desire to exalt instinct and nature, his love of life and vigor, his predilection for the noisome atmosphere of places where the masses spend their social life make him find the odors of dung, stable, fish, butter and cheese, orchards and fruit the most attractive of perfumes. Should his delight in the odors of grocers' shops and, above all, of tanneries (one of the sites most intolerable to delicate senses of smell) be attributed to masochism[50] or to his roots in the sensual life of the masses? It is difficult to decide. "In the depths of the Breuil is the tannery with . . . its sour odor. I adore it, that heady, mustardy, green—if one can say green—odor, like the skins that are hung out in wet weather or left out for the sun to dry their sweat. From as far back as my first visit to the town of Le Puy, and when I returned there later, I smelled the Breuil tannery and guessed its presence—each time one of these factories was within two miles of my route, I sniffed it and turned my grateful nose in that direction."[51]

Jacques Vingtras's olfactory behavior is deeply involved in his revolt—a revolt against a painful past, which the sense of smell, more than any other sense, revives in his memory. Olfaction preserves its discriminative power in the young man's recollection; it bridges the gaps in time.

Vallès uses memories of whole bouquets of odors, yet his virtuosity is not obvious as a literary technique as it is with Zola. In the Pannesac district of Le Puy there was a grocer's "that added to the tranquil odors of the market a muted warm and violent odor exhaled by salted cod, blue cheeses, tallow, fat, and pepper. The cod was dominant, reminding me more than ever of islanders, cabins, glue, and smoked seals."[52] Leaving the town, he retains only the memory of the smells; there is no recall of any balsamic effect such as the psychologists of the *Dictionnaire des sciences médicale* would have liked. "I remember only that I found myself alongside a ditch that smelled unpleasant, and that I walked through a pile of grasses and plants that did not smell good."[53]

The olfactory behavior of the child presages his future involvements.[54] His repulsion to the odor of onions growing in the market gardens on the way out of the town reveals his rejection of the "honest work of the gardens."[55] As an adult, Vingtras associates the pleasant odor of the ink used to print the journal of his revolt with the balsamic scents of the stable. He adopts everything that stifles the bourgeoisie. The Revolution is the countryside and instinct rediscovered. Vallès's love for the Republic exactly parallels his love of dung.[56]

The tradition persisted; recall of smells continued to be the auxiliary of revolt, of pleading the cause of instinct and dissolute childhoods. The tolerance of the hero of Céline's *Mort à credit* for the proximity of excrement and his obsessive terror of everything that implies fecal discipline, the ascending scale of intensity characteristic of the odors of Brooklyn and of its women in Henry Miller's *Tropic of Capricorn,* and the reassuring atmosphere that Gunter Grass's dwarf Matzerath found in his grandmother's skirts bear witness better than anything else to the depths of their wider revolt against the social order.[57]

14 *The Odors*
of Paris

DURING THE SUMMER of 1880, the stench in Paris became so intense that public opinion was roused to revolt. "People greeted each other with only one phrase when they met: 'Can you smell it? What a stink!' It was like a public disaster. The Parisian was driven mad, the prefect harassed, the minister irritated."[1]

A hierarchy of contemporary repulsions emerges from the many writings about the scourge of the stench.[2] They show how out-of-date perception was and how persistent were the old anxieties. Public opinion spontaneously, and wrongly, attributed the scourge to the presence of excrement and other filth in public space: industrial odors were barely considered as a contributing factor.

During October the press exploded. First the Comité d'Hygiène et de Salubrité de la Seine and then the municipal council discussed the problem. The prefect talked of setting up a commission, composed mainly of doctors. The literature provoked by this affair, and especially the experts' report, bore witness to the relative ineffectiveness of the tactics for disinfecting public space.[3] Despite the administration's injunctions, dirt continued to accumulate on public byways. In some districts excrement was still emptied onto the road; children urinated in the street; the clearing of cesspools spread infection day and night. The rapid increase in the number of horses in use in the capital complicated the administration's task. Cesspools grew up around coaching stations; foul vapors from the Champ de Mars spread toward Grenelle and the Gros Caillou. There are repeated references in descriptions of public buildings: the stench in the latrines of La Pitié, as well as in those used by staff in elegant neighborhoods, reached an unprecedented level. As for the dwellings of the masses, the inexhaustible complaints suggest that nothing had changed there since the July Monarchy. But it seems obvious that the Pasteurian revolution made a reappraisal of the old sensitivity and tolerances necessary. The relaxation of censorship that accompanied the triumph of the Republic and the polemics that predominated in the municipal council also favored public debate and vigorous denunciation of abuses.

The Decline of Pre-Pasteurian Mythologies

The uproar about the odors of Paris was abundant proof of the rapid spread of Pasteur's discoveries. In 1880 none of the experts challenged the new theories. The scientific community no longer believed in miasma.[4] Spontaneous generation no longer had any defenders. Once they had become convinced that infectious germs transmitted disease, scientists no longer associated unpleasant odor with the morbific threat. "We can repeat that everything that stinks does not kill, and everything that kills does not stink," declared the conservative Brouardel during the debate.[5] The following year, the *Dictionnaire Dechambre* confirmed the pathogenic discrediting of smell.[6]

Belief in the morbific threat from mud and earth impregnated with putrid materials was also on the wane. Instead, praise was now heaped on their filtering properties, discovered by Schloesing and acknowledged by Pasteur himself. Because soil retained infectious germs, effluvia and exhalations from the earth now attested to the purity of the air. Miquel showed that even emanations from drains "can contain vapors from foul substances; they do not contain microbes."[7] The new discoveries defused the old fear of cemeteries. In 1879 a commission concluded that these places were harmless. "The gases that originate in buried materials in the process of decomposition are always exempt from bacteria," wrote Chardouillet in 1881, while Colin demonstrated that the corpses of buried animals were innocuous.[8]

Experts now also doubted the notion that the exhalations from stagnant water carried miasmas. Miquel demonstrated in 1880 "that water loaded with organic matter that has reached the final stage of putrefaction can be evaporated almost to dryness without a single one of the microgerms that swarm in it being borne off by the vapor. A hundred grams of the water resulting from condensation of this vapor was collected; it had a foul odor like the liquid it came from, but was absolutely pure of all listed miasma."[9]

The discrediting of the pathogenic role of stench hastened the decline of olfaction in clinical diagnosis; the doctor lost his privileged position as an analyst of odors, particularly since as a member of the bourgeoisie he shared the characteristics of his class. The chemical engineer now emerged as the expert in olfaction.

Closed Circuit versus Torrent

Paradoxically, the methods that scientists now recommended to get rid of troublesome stenches were an extension of the tactics used before Pasteur's discoveries. All the experts agreed on the need to seek "the radical ablation of excrements," which were now known to be responsible for typhoid fever.[10] In addition, the new availability of guano from Peru, nitrates from Chile, and, even more, chemical fertilizers militated against the use of human fertilizer.

There were two conflicting types of deodorizing tactics. The reports by the commission of doctors set up to investigate the "odors of Paris" summarize the first type. Based on the isolation technique, its aim was to prevent all contact between the human environment and materials likely to produce germs in large quantities. These tactics did not depend on controlling the flows, but on the principle of creating a vacuum, making systems impervious, and the use of pumps. The commission's plan therefore provided for the installation of cesspools constructed of metal to render them completely hermetic. "On leaving the latrines, the dejecta would be received into absolutely impervious pipes, with metallic linings, without any communication with air or earth. Connected together, these conduits would carry the cesspool material to a place far from the town in which factories would be set up to effect the necessary transformations of these materials . . . circulation [was to be] ensured by lift-and-force pumps, by vacuum, or by any other process."[11] In this scheme the cesspool figured as nothing more than the initial component of a closed circuit with its terminal at the processing factory—unless the system, as Pasteur foresaw, carried the refuse directly into the sea. The plan was to render the waste invisible and odorless and to preserve the population from any contact with it. This approach inspired several achievements: the Liermur system in Belgium and, even better, the pneumatic evacuation network that Berliet set up at Lyons in 1880. Belgrand envisaged this type of solution for Paris in 1861.

These tactics were in opposition to those of the service engineers, who were better informed on foreign achievements and drew strength from the views of the International Congress of Hygiene of 1878. According to them, the commission's proposed solutions posed risks of clogging, required continual handling of a complex system of pumps and taps, and would create an intolerable stench during the inevitable repairs. The engineers' scheme was based on kinetics, not on isolation; it relied not on watertight equipment but on accelerated rhythms. It sought to prevent germs from proliferating by rapidly circulating the excrement. Carried along in the torrent from the drains, the excrement would lose its noxiousness. Their solution would eliminate cesspools, blowoff pipes, cesspool clearance, sulfate of ammonia factories, and dumping grounds; instead there would be "as rapid an evacuation of the waste matter as possible with no stopping anywhere" until it reached the filter beds, where the earth would exert its purifying action.[12]

The engineers' plan was based on English discoveries made before Pasteur's. English experts had proved that excrement presented no danger and did not emit very unpleasant odors until the second day; the time lag could be increased by movement. English books about river pollution showed that water from drains was innocuous. On the Continent, Freycinet gave these daring theories the authority of his support.[13]

The great network of London sewers, built at the beginning of the 1860s, was the result of these findings. The system had been adopted in Brussels, Frankfurt-am-Main, and Danzig. It was being implemented in Berlin; a commission headed by Rudolf Virchow had just decided on its superiority. In both England and the United States the issue was no longer the merits and demerits of sewerage but whether a dual network, one for rainwater and one for sewage, was essential.

The immediate removal of excreta, drowned in a stream of water, was obviously the most effective deodorization technique for both public and private space. Protracted resistance by French administrators explains more than anything else the persistence of urban stenches.

Stagnation or Dilution The backwardness of French administrators was an extremely significant historical fact. There was considerable congruence between the conflicting schemes of the sanitary reformers and the prevailing systems of social perception. Both centered on the conflict between the tactics of stagnation and isolation on the one hand and those of movement and dilution on the other. Brouardel, the most determined supporter of the hermetic system, was also the most outspoken champion of controlling prostitution by licensing brothels. In both instances he relied on the interests of property; he was the self-appointed spokesman for the madams of brothels and the cesspool-clearance companies, traffickers in lost flesh and in accumulated excrement. In both cases he was outflanked by the reformers.

The champions of movement and dilution stressed the egalitarian virtues of their scheme. It did in fact imply the availability of water for everyone. Its advocates launched a long diatribe against landlords who refused to pay water rates. While eau de cologne and scented soap became widespread, the engineers—Republicans—of the Paris Highways Department demanded equal treatment for the excrement of rich and poor alike. To support their case, they documented differences in the handling of excrement among various social classes and districts.

Pasteur's theories inevitably influenced social perceptions and tactics. The discovery of the microbial germ required a revision of the epidemiology initiated by Villermé. The morbific danger became more diffuse, less easily perceptible, and also more disturbing. All water had to be "suspect," declared Marié-Davy;[14] and it is tempting to add: all individuals too. The universal presence of microbes increased the need for a comprehensive strategy embracing all strata of the population, as scientists were well aware. "Shared life in a large town makes us all interdependent on one another . . . These organisms [microbes] are diffused in the air outside and penetrate everywhere, into our apartments, into our lungs, into our drink, into our food . . . The hygiene of a town can never be ensured so long as it continues to be neglected in its poor districts."[15] These beliefs required new types of alarm signals and a revision of social hygiene tactics.[16]

Nevertheless attitudes were by no means modern by our standards. The alliance between germs and dirtiness—now identified with filth and dust—remained unchallenged. There were fifty to sixty times more microbes in the poor man's dwelling than in air from the most evil-smelling sewer, declared Marié-Davy in 1882.[17] Stench was no longer morbific, but it signaled the presence of disease. The masses had lost their monopoly on infection, but they remained the greatest threat.

Bourgeois families were now obsessed by fear of degeneration. Microbes thrived and proliferated in the blood of the masses; they flourished amid vice and dirtiness, in streets, slums, and servants' quarters. From contact with the proletariat the bourgeois ran the risk not only of contagion but also of biological mutation: the virulent germ, rising out of the social slime, might well be transmuted into a hereditary taint in the blood. In that event, his posterity was jeopardized, his genetic patrimony was threatened with corruption.[18]

Thus although the danger had become more diffuse and harder to detect, the demand for social segregation persisted, only in a more subtle form. The revision of guidelines, manifest in the new control of prostitution, was only the most obvious example of an immensely significant strategic readaptation of the greatest significance—one based on the systematic medical examination of the population as a whole.

The persistence of stench in Paris bore witness to the slow pace of the evolution in administrative practices. Until the eve of World War I, although a sewage system had been approved in 1889 and the aqueduct of Achères completed in 1895, the capital continued to stink in the summer. Every year Adam, chief inspector of public buildings, drew attention to the scourge; he even tried to list the worst-smelling days. Nothing was done about it even when the service was reorganized in 1897.

Sporadic campaigns like those conducted against prostitution attempted to rouse public opinion against official incompetence. During the summer of 1911 the crisis exploded. The smell stifled pedestrians, particularly in the evening; experts claimed it was a stench "of waxing, of heated organic material."[19] This time, thanks to Verneuil, the culprits were ascertained: they were the super-phosphates factories in the northern suburbs.[20] The industrial belt forced its guilty stench upon the capital as the abominable Montfau-con had in the past. Industry had replaced excrement in the hierarchy of repulsion. A new ecological awareness took its first, rudimentary shape.

Conclusion

THE MEN AND WOMEN of the nineteenth century muffled history with the clamorings of
their desire. Democrats dreamed of "la Belle
République"; Michelet invented "the People"; socialists designed the happiness of man-
kind; positivists preached the education of the masses. Meanwhile,
however, other dialogues were taking place at a more fundamental
level; heavy animal scents and fleeting perfumes spoke of repulsion
and disgust, sympathy and seduction.

Despite Lucien Febvre's injunctions, historians have neglected
these documents of the senses. The sense of smell was discredited.
According to Buffon, it was the sense of animality. Kant excluded
it from aesthetics. Physiologists later regarded it as a simple residue
of evolution. Freud assigned it to anality. Thus discourse on odors
was interdicted. But the perceptual revolution, precursor of our
odorless environment, can no longer be suppressed.

The decisive action was played out between 1750 and 1880, in
the heyday of the pre-Pasteurian mythologies. The history of science,
teleological in form, concerned solely with the progress of truth,
scornful of the historical consequences of error, has hitherto neglected this drama. In about 1750, the work by Pringle and MacBride
on putrid substances, the rise of "pneumatic" chemistry, and the

phantasm of urban pathology suggested a new cause for disquiet. Excrement, mud, ooze, and corpses provoked panic. This anxiety, flowing from the peak of the social pyramid, sharpened intolerance of stench. It fell to the sense of smell to destroy the confused issue of the putrid, to detect miasma in order to exorcise the malodorous threat.

Contemporary scientists, incomparable observers of odors, offered a fragmented, olfactory image of the town; they were obsessed with pestilential foci of epidemics. To escape this swamp of effluvia, the elite fled from social emanations and took refuge in fragrant meadows. There, the jonquil spoke to them of their "I," inspired the poetry of the "nevermore," and revealed the harmony between their being and the world.

Musk, a waste product originating in the putrid guts of the musk deer, began to arouse repugnance. There were threats amassed in it as well. Its evocations of female odor became intolerable. The new fashion for delicate scents expelled it from the court while public health tacticians attempted to purify and deodorize public space.

After the Revolution, with its fascination with corpses and scorn for vegetable scents, the return of musk took on symbolic value. Sprinkled with eau de cologne, drenched in vapors from animal perfumes, the imperial couple broke with rose water. The Restoration also expressed itself in terms of smell. In this respect the faubourg St.-Germain evinced the morbid sensitivity of a chlorotic girl. Vegetable perfumes reimposed their delicacy; their function was to dampen female impulses and to signal a new system of control.

At the same time, fear of the obtrusive presence of a dangerous human swamp replaced the obsession with carrion and ooze, swarming with noxious miasmas. In the hierarchy of anxieties, there was a shift from the vital to the social: instinct, animality, and organic stench became traits of the masses. Repugnance to smell now focused on the poor man's hovel and latrines, the peasant's dung, the greasy and fetid sweat impregnating the worker's skin, rather than on the oppressive vapor of the putrid crowd in general. Flaubert could not sleep for having breathed the odor from the proletarian omnibus; Adolphe Blanqui recoiled appalled from the mephitic blast exhaled by the "ditches of men" where the Lille weavers crowded.

Thenceforth these more discriminatory maneuvers of the sense of smell were required to strengthen what were perceived as increasingly complex hierarchies. Repelled by the secretions of poverty, the bourgeoisie became alert to subtle bodily messages, the go-be-tweens of seduction. Their growing importance compensated for the ban on contact.

Far from the odor of the masses, the bourgeoisie set out, albeit clumsily, to purify the breath of the house: rooms had to be aired after the maid had stayed in them for an extended period, after a peasant woman had called, or after a workers' delegation had passed through. Latrines, kitchens, and dressing rooms gradually ceased to give off their intrusive scents. Lavoisier's chemistry made it possible to define precise norms of ventilation. Salons and boudoirs became the settings for a new and skillful arrangement of scents. Trouble-some odors would no longer disturb the bedroom, temple of private life and intimacy.

After Novalis, a silent dialogue, woven of symbols, was initiated between flowers and the young girl or woman. Vegetable perfume, issuing a delicate invitation, refined the interchange. It permitted the expression of desire and female solicitation but also the maintenance of physical distance. The fragrant alleys of the bourgeois garden revived amorous dialogue. The lover went there to taste intoxication in the mode of anticipation, whereas the plebeian male, overwhelmed by genetic instinct, seized his conquest. Patient breathing alongside the loved one, a skillful delaying preliminary, guaranteed constancy of desire. Recollection of the smell of the other person's body kept passion alive and nourished desire for the absent one; it was an incentive to the neurotic collection of mementos.

Outside, the deodorization of roads, spurred by the use of chlo-rides, the utilitarian approach to refuse, and the new intolerance of industrial pollution, no longer satisfied officials' ambitions. They now launched their sanitizing enthusiasms on the dirt of the wretched poor. They launched inspections of insalubrious dwellings, schools, barracks, and bathhouses in sports clubs. But it was a long time before bodily hygiene achieved any decisive success among the masses. For the time being, efforts concentrated on the appearance of cleanliness and particularly on fecal discipline. In this climate of opinion, deo-dorization encountered muffled resistance. The old patterns of per-ception and appraisal persisted; habit kept alive nostalgia for free organic manifestations.

It is from the sense of smell, rather than from the other senses, that we gain the fullest picture of the great dream of disinfection and of the new intolerances, of the implacable return of excrement, the cesspool epic, the sacralization of woman, the system of vegetable symbols. It permits a new interpretation of the rise of narcissism, the retreat into private space, the destruction of primitive comfort, the intolerance of promiscuity. Distinctions and disagreements were deeply rooted in two opposed conceptions of air, dirt, and excrement; they were expressed in the antithetic conduct of the rhythms and fragrances of desire. Only an absence of smell in a deodorized environment—our own—achieved resolution of the conflict.

This episode in the history of disgust, affinities, and purification, spanning the nineteenth century, revolutionized social perceptions and symbolic references. Without knowledge of that history, we can neither measure the visceral depths to which the nineteenth-century social conflicts reached nor explain the present vitality of the ecological dream.

Social history, respectful toward the humble but indifferent for too long to the expression of emotions, must no longer suppress people's elementary reactions, however sordid, on the pretext that the delirious anthropology of the Darwinian period has perverted their analysis.

NOTES

INDEX

NOTES

Introduction

1. J.-N. Hallé, "Procès-verbal de la visite faite le long des deux rives de la rivière Seine, depuis le pont-Neuf jusqu'à la Rappée et la Garre, le 14 février 1790," *Histoire et Mémoires de la Société Royale de Médecine*, 10 (1789), lxxxvi.
2. J.-N. Hallé, *Recherches sur la nature et les effets du méphitisme des fosses d'aisances* (1785), 57–58.
3. J.-N. Hallé, "Air des hôpitaux de terre et de mer," in *Encyclopédie méthodique. Médecine* (1787), 571.
4. On the pleasures of sight in the eighteenth century, see Mona Ozouf, "L'Image de la ville chez Claude-Nicolas Ledoux," *Annales. Economies, Sociétés, Civilisations*, 21 (November–December 1966), 1276.
5. Lucien Febvre, *Le Problème de l'incroyance au XVIᵉ siècle* (Paris, 1942).
6. In his *Introduction à la France moderne. Essai de psychologie historique, 1500–1640* (Paris, 1961), Robert Mandrou, influenced by Lucien Febvre, devotes a long chapter to the history of perception at the dawn of modern times; as far as I know this is the only attempt at a synthesis of the subject.

   Since the publication of Pierre Francastel's work, the historical analysis of sight has inspired numerous books, most recently those by Michael Baxandall. Number 40 of *Actes de la Recherche en Sciences Sociales* (1981) is devoted entirely to this aspect of the sociology of perception. In *La Fantasmagorie* (Paris, 1982), a masterly work devoted to the study of the mirror image and the transfiguration of the perceptive universe in the literature of fantasy, Max Milner analyzes the bonds that, according to Kant, link sensory history and the inquiry into identity.

   As early as 1967, Jean-Paul Aron's *Essai sur la sensibilité alimentaire à Paris au XIXᵉ siècle* (Paris) inaugurated a long series of works devoted to the history of taste. The Institut Français du Gout periodically endeavors to bring together at Tours all researchers into the human sciences concerned with psychosociology and the history of eating behavior. However, very few of these studies concern the gustative sensation, the poverty of which is well known: it is in fact the sense of smell that contributes the refinement of flavors.

   Mention must also be made of the interesting book by Ruth Winter, a journalist for the *Los Angeles Times: Scent Talk among Animals* (New York, 1977). It contains a copious bibliography of recent works in physiology and experimental psychology and, notably, references to books by J. Le Magnen and by A. Holley, French experts in these aspects of osmology. The aesthetics of the sense of smell is the subject of Edmond Roudnitska's remarkable

L'Esthétique en question (Paris, 1977), which includes an interesting study of Kant's rejection of the sense of smell.

Finally, mention should be made of the whole body of work by Peter Reinhart Gleichmann. For some years he has been extending the research of Norbert Elias and studying the interrelationships between the change in emotions, the transformation of the images of the body, and the techniques of social control that the construction of cleansing systems reveals. Especially relevant to our purposes are his articles on the integration of the physiological functions in the domestic sphere and the extent of the chain reactions engendered by this domestication; see, for example, "Des villes propres et sans odeur," *Urbi,* April 1982. However, his primary focus is central Europe between 1866 and 1930; he says nothing about the pre-Pasteurian mythologies and minimizes the importance of the period studied here. In the same field see also Dominique Laporte, *Histoire de la merde* (Paris, 1979).

7. See Jean Ehrard, *L'Idée de nature en France dans la première moitié du XVIIIe siècle* (Paris, 1963), 676.

8. Ibid., p. 685.

9. A survey of these episodes in Enlightenment philosophy is beyond the scope of this book. Claire Salomon-Bayet, *L'Institution de la science et l'expérience du vivant* (Paris, 1978), 204 ff., has successfully analyzed how scholars use the observations of the *homo ferus,* the philosophical fiction (Condillac's statue), the experimental fictions (the healed blind man of Maupertuis), or unforeseen accidents (Rousseau's fall during his second promenade) to try to solve the problems posed by empirical knowledge.

10. Jacques Guillerme, "Le Malsain et l'économie de la nature," *XVIIIe Siècle,* 9 (1977), 61.

11. "All the diversities of flavors, odors, sounds, colors, in a word, all our sensations are only the action of God upon us, diversified according to our needs"; Pluche, *Spectacle de la nature,* vol. 4 (1739), 162.

12. Febvre, *Le Problème de l'incroyance,* 461–472.

13. Emphasized by Locke, *An Essay Concerning Human Understanding* (London, 1947), 36 (1st ed. 1755).

14. Robert Boyle, *The General History of the Air* (London, 1692), had noted that musk, despite the strong odors it emitted, lost nothing, or almost nothing, of its substance. Albrecht von Haller, *Elementa physiologiae corporis humani,* vol. 5 (Lausanne, 1763), 157, kept papers perfumed by a single grain of ambergris for more than forty years with no diminution in their odor. Hermann Boerhaave propounded as many observations confirming the *spiritus rector* theory. In his view, far from being the emanation of corpuscles separated from the smelling body, odor was a subtle fluid, "a volatile being, very fleeting, very expansible, weightless, completely invisible, inaccessible to the senses were it not for the olfactory membrane"; quoted in Hippolyte Cloquet, *Osphrésiologie ou traité des odeurs,* 2d ed. (1821), 39–40. For the majority of scholars that guiding spirit, called aroma at the end of the eighteenth century, was oily in nature. It seemed obvious, however, that it did not assume the same form everywhere, and Pierre Joseph Maquer, one of the most eminent chemists of the day, strove to catalog its various manifestations.

It was precisely this variety that eventually discredited Boerhaave's theory. Since aroma perpetually proved different from itself, its existence as a principle could no longer be sustained. This at least is what Nicolas Le Cat (*Traité des sensations et des passions en général et des sens en particulier,* vol. 2 [1767], 234) and the chevalier Louis de Jaucourt ("Odorat," in the 1765 Encyclopédie) already thought. Although the corpuscular theory, already formulated by Theophrastus and approved by the Cartesians, remained hypothetical until Fourcroy and Berthollet proved it well founded, a number of Hallé's contemporaries thought that bodies emitted several particles of smell that formed part of their substance.

15. Especially for Buffon.
16. See Condillac's view of the role of language; Ehrard, *L'Idée de nature,* 686.
17. Haller, quoted in "Odorat," *Encyclopédie,* supplement (1777).
18. Père du Tertre, *Histoire naturelle et morale des îles Antilles . . .* (1658); Père Joseph François Lafitau, *Moeurs des sauvages américains . . .* (1724); Alexander von Humboldt, *Essai politique sur le royaume de la Nouvelle-Espagne* (1811).
19. Notably Samuel Thomas von Soemmerring and Johann Friedrich Blumenbach.
20. "There have been observations," wrote Haller again in 1777, "of a child raised in a wilderness sniffing the grass as a sheep would, and choosing by the odor the piece he would like to eat. Having been returned to society and become accustomed to different foods, he lost this trait"; quoted in "Odorat."
21. See Le Cat, *Traité de sensations,* 230.
22. An opinion also found in Haller, *Eléments de physiologie,* 2 vols. (1769), 2:33.
23. Jaucourt, "Odorat."
24. Haller, "Odorat."
25. Rousseau, *Emile* (Paris, 1966), 200–201; odors "exert an influence not so much by what they give forth as by what they hold in abeyance."
26. Jaucourt, "Odorat": "There is a mysterious rapport between the vital principle and fragrant bodies."
27. "I began to see calmly and to hear effortlessly when a light, fresh breeze brought me scents that caused me to blossom inwardly and gave me a feeling of love for myself," declares the first man in Buffon's account; *De l'homme* (Paris, 1971), 215.

1. Air and the Threat of the Putrid

1. For example, Boissier de Sauvages, winner of the Dijon Academy competition on this subject in 1753, remained loyal to the mechanist conception of the air, namely, that it was composed of small spheres or molecules separated by hollow interstices from which other substances seeped out; François Boissier de Sauvages, *Dissertation où l'on recherche comment l'air, suivant ses différentes qualités, agit sur le corps humain* (1754). In the previous century, Boerhaave had regarded the air as a simple tool, an intermediary not involved in chemical exchanges.
2. "That is why," Malouin wrote in 1755, "the same foods are digested differently, according to the differences in the air breathed"; consequently, digestion was

better in the country than in the town. Paul-Jacques Malouin, *Chimie médicinale,* vol. 1, p. 54.

3. On the importance that the concept of fiber assumed in the eighteenth century, see Jean-Marie Alliaume, "Anatomie des discours de réforme," in Comité de la Recherche et du Développement en Architecture, *Politiques de l'habitat (1800–1850)* (Paris, 1977), 150.

4. Jean Ehrard, *L'Idée de nature,* has written some illuminating articles on this subject.

5. Men of letters, Paul Victor de Sèze noted, were well aware that the morning air "gives a strange disposition to study"; *Recherches physiologiques et philosophiques sur la sensibilité ou la vie animale* (1786), 241.

6. On this subject see the fine article by Owen Hannaway and Caroline Hannaway, "La Fermeture du cimetière des Innocents," *XVIIIe siècle,* 9 (1977), 181–191.

7. In his eyes, the electric fluid constituted the neural fluid itself—which was tantamount to discounting the theory of animal spirits.

8. See Ehrard, *L'Idée de nature,* 701 ff.

9. Boyle, *General History of the Air.* See also John Arbuthnot, *An Essay Concerning the Effects of Air on Human Bodies* (1733), especially 92 ff.

10. For Hippocrates' work and its significance, see Robert Joly, *Hippocrate, médecine grecque* (Paris, 1964), especially "Des airs, des eaux, des lieux," 75 ff. The influence that, depending on school, Greek doctors attributed to the air was extremely complex; see Jeanne Ducatillon, *Polémiques dans la collection hippocratique* (Thesis Paris IV, 1977), 105 ff. The Hippocratic treatises, by subordinating medicine to knowledge of the human body, moved away from the medical theories of the philosophers, who claimed to explain all diseases by the same cause and adopted a cosmological viewpoint giving greater emphasis to the winds than did the doctors of the school of Kos. See the analyses of the treatise "Des vents" by Joly (*Hippocrate,* 25–33) and Ducatillon *(Polémiques).*

  Quite recently, however, Antoine Thivel (*Cnide et Cos? Essai sur les doctrines médicales dans la collection hippocratique* [Paris, 1981]) has questioned the legitimacy of this distinction between the two schools. On the medicine of constitutions, see Jean-Paul Desaive, Jean-Pierre Goubert, Emmanuel Le Roy Ladurie, and Jean Meyer, . . . *Médecins, climats et épidémies à la fin du XVIIIe siècle* (Paris, 1972).

11. See Dr. Pierre Thouvenel, *Mémoire chimique et médicinal sur la nature, les usages et les effets de l'air, des aliments et les médicaments, relativement à l'économie animale* (1780).

12. Arbuthnot, *Essay Concerning Effects of Air,* 208–209.

13. Thouvenel, *Mémoire chimique,* 27. "The air," he wrote belatedly but always from this perspective, "must be neither too virginal nor too spirituous, neither too vapid nor too keen, neither too heavy nor too dull, not too concentrated nor too solvent, not too diluted nor too stale, not too exciting nor too nourishing, neither too septic nor too antiseptic, not too drying nor too moist, nor too relaxing, etc." (p. 24).

14. Arbuthnot, *Essay Concerning Effects of Air,* 213–215.

15. See Jean Ehrard, "Opinions médicales en France au XVIIIe siècle: La peste

et l'idée de contagion," *Annales. Economies, Sociétés, Civilisations*, 12 (January–March 1957), 46–59.

16. Guillerme, "Le Malsain," 61–72.
17. Karl Wilhelm Scheele's work summarized this compulsive labor very well, and, better still, Jean-Godefroi Léonhardy's *Supplément au traite chimique de l'air et du feu de M. Scheele* and *Tableau abrégé des nouvelles découvertes sur les diverses espèces d'air (1785)*.
18. Joseph Priestley, *Observations on Air* (London, 1774).
19. See Guillerme, "Le Malsain," 63.
20. Ibid., p. 61.
21. Pierre Darmon, *Le Mythe de la procréation à l'âge baroque* (Paris, 1977).
22. Thouvenel, *Mémoire chimique*, 13.
23. In his *Historia Naturalis*, especially in *Historia Vitae et Mortis* (1623); for the history of research into decomposition, see Jacques-Joseph de Gardane, *Essais sur la putréfaction des humeurs animales (1769)*.
24. Gardane, *Essais sur la putréfaction*, V.
25. In Greek antiquity, solar and imputrescible aromatics, with myrrh as their archetype, formed the antithesis to humid and putrescible vegetation, symbolized by lettuce. See Marcel Détienne, *Les Jardins d'Adonis. La mythologie des aromates en Grèce* (Paris, 1972).
26. John Pringle, "Some experiments on substances resisting putrefaction," *Philosophical Transactions*, 47 (1750), 480–488, 525–534. David MacBride, *Essais d'expériences* (1766).
27. The treatises by Barthélemy-Camille Boissieu, Toussaint Bordenave, and Guillaume-Lambert Godart are published collectively as *Dissertations sur les antiseptiques* . . . (Dijon, 1769).
28. Quoted in Gardane, *Essais sur la putréfaction*, 121.
29. Robert Mauzi, *L'Idée du bonheur au XVIIIᵉ siècle* (Paris, 1960), 273 ff.
30. Madame Thiroux d'Arconville, *Essai pour servir à l'histoire de la putréfaction* (1766).
31. Godart, *Dissertations sur les antiseptiques*, 253–258.
32. Quoted in Gardane, *Essais sur la putréfaction*, 220.
33. Ibid., p. 124.
34. Guillerme, "Le Malsain," 61.
35. Ehrard, "Opinions médicales," studies the origin and evolution of the theory of miasmas and its initial link with the corpuscular theories that emerged from Boyle's work. Ehrard makes a distinction among this theory of miasmas, that of leavens, and that of worms or insects.
36. Guillerme, "Le Malsain," 63.
37. John Cowper Powys, *Morwyn*. Robert Favre, *La Mort dans la littérature et la pensée française au siècle des Lumières* (Paris, 1978), 403, recalls Chamfort's version of Saint Theresa de Avila's definition of hell as "the place where it stinks and where there is no love."
38. Compare the conviction that obsessed the Romantics, namely that death was necessary for the birth of a new world. Examples are the death of Gauvin and Cimourdin in Hugo's *Quatre-vingt-treize* and, much earlier, Novalis's *Heinrich von Ofterdingen*.
39. Guillerme, "Le Malsain," 62.

2. *The Extremes of Olfactory Vigilance*

1. Ehrard, *L'Idée de nature*, 710.
2. Boissier de Sauvages, *Dissertation*, 51.
3. Ibid.
4. This is the title of Becher's book, which was published in Frankfurt in 1669.
5. This idea of a counterbalance correcting mephitism underlies Arbuthnot's *Essay Concerning Effects of Air.*
6. See Boyle, *General History of the Air.*
7. Bernardino Ramazzini, *De morbis artificum diatriba* (Padua, 1713), trans. A. F. de Fourcroy (1777), 533, 327, 534.
8. See Chapter 9, pp. 155–156.
9. M. de Chamseru, "Recherches sur la nyctalopia," *Histoire et Mémoires de la Société Royale de Médecine*, 8 (1786), 167 ff.
10. J.-B. Théodore Baumes, *Mémoire . . . sur la question: Peut-on déterminer par l'observation quelles sont les maladies qui résultent des émanations des eaux stagnantes . . .* (1789), 234.
11. Ibid., p. 165. In 1815 Etienne Tourtelle still echoed this complaint; *Eléments d'hygiène*, vol. 1, p. 277.
12. Paul Savi, "Considérations sur l'insalubrité de l'air dans les Maremmes," *Annales de Chimie et de Physique*, 3d. ser., 1 (1841), 347.
13. Tourtelle, *Eléments d'hygiène*, 278.
14. An obscure subject, earlier touched on by Jean Roger in *Les Sciences de la vie dans la pensée française du XVIIIe siècle* (Paris, 1963), 642–647. Jean-Baptiste Robinet *(De la nature)* became the exponent of this theory of universal vitality.
15. Michel-Augustin Thouret, *Rapport sur la voirie de Montfaucon*, 13 (read November 11, 1788, to the Société Royale de Médecine).
16. "Rapport fait à l'Académie Royale des Sciences le 17 mars 1780 par MM. Duhamel, de Montigny, Le Roy, Tenon, Tillet, et Lavoisier, rapporteur," *Mémoires de l'Académie des Sciences (1780)*, in Antoine Laurent Lavoisier, *Oeuvres* (1865), 3:493; the italics are mine.
17. For the symbolic value of the deepest dungeon and its role in preserving messages from the past, see, for example, Victor Hugo, especially *Quatrevingt-treize* and *L'Homme qui rit.*
18. See Boissier de Sauvages, *Dissertation*, 54.
19. Bruno Fortier, "La Politique de l'espace parisien," in *La Politique de l'espace parisien à la fin de l'Ancien Régime*, ed. Bruno Fortier (Paris, 1975), 32.
20. Louis-Sébastien Mercier, *Tableau de Paris*, 12 vols. (1782–1788), 1:21.
21. Fortier, "La Politique de l'espace parisien," 116–125; Favre, *La Mort dans la littérature*, 398.
22. See Bachelard, *La Terre et les rêveries de la volonté* (Paris, 1948), 129 ff.; this concern with muddy material concealed an ambivalence that has provided psychoanalysts with material for copious dissertations.
23. See Pierre Chauvet, *Essai sur la propreté de Paris* (1797), 24, and especially Mercier, *Tableau de Paris*, 1:213, and J.-H. Ronesse, *Vues sur la propreté des rues de Paris* (1782), 14. The precision with which the latter two analyze the

mud of the streets of Paris reveals the importance they attached to the substances. The texts quoted by Pierre Pierrard on the muds of Lille show the same analytic precision; *La Vie ouvrière à Lille sous le Second Empire* (Paris, 1965).

24. Alexandre Parent-Duchâtelet, "Essai sur les cloaques ou égouts de la ville de Paris" (1824), in *Hygiène publique,* 2 vols. (1836), 1:219–220.

25. Michel Eugène Chevreul, "Mémoire sur plusieurs réactions chimiques qui intéressent l'hygiène des cités populeuses," *Annales d'Hygiène Publique et de Médecine Légale,* 50 (1853), 15 (read November 9 and 16, 1846).

26. Ibid., pp. 36, 38.

27. Ibid., p. 17.

28. P.-A. Piorry, *Des habitations et de l'influence de leurs dispositions sur l'homme en santé et en maladie* (1838), 49.

29. Mercier, *Tableau de Paris,* 4:218.

30. John Howard, *The State of the Prisons,* 3d ed. (London, 1784), 88.

31. Philippe Passot, *Des logements insalubres, de leur influence et de leur assainissement* (1851). On this subject, Passot quotes (p. 24) Francis Devay's *L'Hygiène des familles.*

32. Quoted in Passot, *Des logements insalubres,* 25.

33. Ibid.

34. Mathieu Géraud, *Essai sur la suppression des fosses d'aisances et de toute espèce de voirie, sur la manière de convertir en combustibles les substances qu'on y renferme* (Amsterdam, 1786), 34.

35. James Lind, *An Essay on the most effectual Means of preserving the Health of Seamen* (London, 1757), 17; Henri Louis Duhamel-Dumonceau, *Moyens de conserver la santé aux équipages des vaisseaux; avec la manière de purifier l'air des salles des hôpitaux* (Paris, 1759), 131.

36. Howard, *State of Prisons,* 346–348.

37. John Howard, *An Account of the Principal Lazarettos in Europe* (Warrington, 1789), 144.

38. There were similar complaints about wool's capacity for impregnation.

39. Chauvet, *Essai sur la propreté de Paris,* 17.

40. Mercier, *Tableau de Paris,* 7:226.

41. There is a long digression on this subject in Alfred Franklin, *La Vie privée d'autrefois,* vol. 7, *L'Hygiène* (Paris, 1900), 153 ff.

42. On the general ineffectiveness of the Paris police, see Arlette Farge, *Vivre dans la rue à Paris au XVIIIᵉ siècle* (Paris, 1979), 193 ff., especially p. 209.

43. In fact these craftsmen used stagnant urine; see Ramazzini, *De morbis artificum,* 149.

44. Mercier, *Tableau de Paris,* 11:54.

45. Chauvet, *Essai sur la propreté de Paris,* 18.

46. La Morandière (1764), quoted by Dr. Augustin Cabanès, *Moeurs intimes du passé* (Paris, 1908), 382.

47. Arthur Young, *Travels during the years 1787, 1788 and 1789 . . . of the Kingdom of France* (London, 1792), 160.

48. John Pringle, *Observations on the Diseases of the Army* (1752), 300. Pringle based his conclusions on the experiments conducted by Homberg as early as 1711.

49. Géraud, *Essai sur la suppression des fosses,* 38.

50. Laborie, A.-A. Cadet the younger, and A.-A. Parmentier, *Observations sur les fosses d'aisances et moyens de prévenir les inconvénients de leur vidange* (1778), 106.

51. Chauvet, *Essai sur la propreté de Paris,* 38.

52. Mercier, *Tableau de Paris,* 11:55.

53. Thouret, *Rapport sur la voirie,* 15.

54. Géraud, *Essai sur la suppression des fosses,* 66.

55. Ibid., p. 96.

56. Alexandre Parent-Duchâtelet, *Rapport sur les améliorations à introduire dans les fosses d'aisances,* reprinted in *Hygiène publique,* 2:350.

57. Voltaire, "Déjection," in *Dictionnaire philosophique* (Geneva, 1764).

58. Mercier, *Tableau de Paris,* 10:250.

59. Pierre J.-B. Nougaret and J.-H. Marchand, *Le Vidangeur sensible* (1777). The article on nauseas that gave rise to the idea of staging this play—with the professed aim of combatting the "affectation" of disgust (p. xiv)—attests to the fascination with excrement and to the new sensitivity.

60. Hallé, *Recherches sur la nature du méphitisme,* 77–81. Precise analyses can also be found in the books cited by Laborie and Thouret.

61. Thouret, *Rapport sur la voirie,* 21.

62. Mercier, *Tableau de Paris,* 7:229: "The excrement of the masses with their varied shapes is constantly before the eyes of duchesses, marquesses, and princesses." Not until the nineteenth century was there an attempt to identify the odor of excrement only with the poor, and this happened on another count.

63. See especially Philippe Ariès, *L'Homme devant la mort* (Paris, 1978); Pierre Chaunu, *La Mort à Paris, XVIe, XVIIe, XVIIIe siècles* (Paris, 1978); Pascal Hintermeyer, *Politiques de la mort* (Paris, 1981); and François Lebrun, *Les Hommes et la mort en Anjou aux XVIIe et XVIIIe siècles* (Paris, 1975).

64. Abbé Charles Porée, *Lettres sur la sépulture dans les églises* (Caen, 1745).

65. Henri Haguenot, *Mémoire sur les dangers des inhumations* (1744).

66. Félix Vicq d'Azyr, *Essai sur les lieux et les dangers des sépultures* (1778), cxxxi.

67. Dr. Jacques de Horne, *Mémoire sur quelques objets qui intéressent plus particulièrement la salubrité de la ville de Paris* (1788), 4.

68. See A.-A. Cadet de Vaux, *Mémoire historique et physique sur le cimetière des Innocents* (1781).

69. Charles Londe, *Nouveaux Eléments d'hygiène,* 2 vols. (1838), 2:348.

70. François-Emmanuel Fodéré, *Traité de médecine légale et d'hygiène publique ou de police de santé . . . ,* 6 vols. (1813), 5:302.

71. Such analyses led Pierre-Toussaint Navier to construct a theory that morbific rays rose from corpses; *Sur les dangers des exhumations précipitées et sur les abus des inhumations dans les églises* (1775).

72. See Louis Jean Marie Daubenton et al., *Rapport des mémoires et projets pour éloigner les tueries de l'interieur de Paris.* At that time sixteen slaughterhouses operated in the open air along the rue St.-Martin from the rue Au Maire to the rue Montmorency, and six others along adjacent streets.

73. De Horne, *Mémoire,* 11.

74. Thouret, *Rapport sur la voirie*, 28. Stench was a basic element of urban pathology; see Emmanuel Le Roy Ladurie, *Histoire de la France urbaine*, vol. 3 (Paris, 1981), 292 ff.
75. Mercier, quoted on p. 54.
76. See especially M. F.-B. Ramel, *De l'influence des marais et des étangs sur la santé de l'homme* (Marseilles, year X) (first published in 1784 in the *Journal de Médecine*).
77. Malouin, *Chimie médicinale*, 62.
78. Duhamel-Dumonceau, *Moyens de conserver la santé*, 40.
79. Abbé Pierre Bertholon, *De la salubrité de l'air des villes et en particulier des moyens de la procurer* (Montpellier, 1786), 6, 7.
80. Moreover, the word *pollution* did not then have the meaning we give it.
81. Joseph Raulin (1766), quoted in Ramel, *De l'influence des marais*, 63.
82. Fodéré, *Traité de médecine légale*, 5:168. See Jean-Baptiste Monfalcon, *Histoire des marais* (1824), 32. This work also gives a well-documented synthesis of the history of theories devoted to the "nature of emanations from swamps" (pp. 69–78). The quays of the Charente were the subject of a considerable literature at the beginning of the nineteenth century. See Alain Corbin, "Progrès de l'économie maraîchine," in *Histoire du Poitou, du Limousin et des pays charentais*, ed. E. R. Labande (Toulouse, 1976), 391 ff. and bibliography, 413–414.
83. Baumes, *Mémoire*, 99.
84. Jan Ingenhousz considered that the gases were phlogisticated, septic, and putrid all at the same time; *Expériences sur les végétaux, spécialement sur la propriété qu'ils possèdent à un haut degré soit d'améliorer l'air quand ils sont au soleil, soit de le corrompre la nuit ou lorsqu'ils sont à l'ombre* (Paris, 1787), 167 (1st ed. London, 1779). To contemporary scientists this was confirmation that swampy emanations were an aggregation of every menace.
85. Baumes, *Mémoire*, 7.
86. Ibid.
87. Fodéré, *Traité de médecine légale*, 5:164 ff.
88. Baumes, *Mémoire*, 196.

3. Social Emanations

1. Vitalist doctrine held that not all life processes could be explained by the ordinary laws of physics and chemistry. As early as 1786, de Sèze (*Recherches physiologiques*, 85) asserted that it was Bordeu and Anne-Charles Lorry who, with Paul Joseph Barthez, ensured the defeat of the mechanistic perspective, with its springs, pumps, and levers.
2. Théophile de Bordeu, *Recherches sur les maladies chroniques*, vol. 1 (1775), 378, 379, 383.
3. Dr. Jean-Joseph de Brieude, "Mémoire sur les odeurs que nous exhalons, considérées comme signes de la santé et des maladies," *Histoire et Mémoires de la Société Royale de Médecine*, 10 (1789); Julien-Joseph Virey, "Des odeurs que répandent les animaux vivants," *Recueil Périodique de la Société de Médecine de Paris*, 8 (year VIII), 161 ff. and 241 ff.; Augustin-Jacob Landré-Beauvais,

"Des signes tirés des odeurs," in *Séméiotique ou traité des signes des maladies,* 2d ed. (1815), 419–432.

4. Isis Edmund Charles Falize, "Quelle est la valeur des signes fournis par l'odeur de la bouche?" in *Questions sur diverses branches des sciences médicales* (Thesis, Paris, April 12, 1839).

5. Dr. Ernest Monin, *Les Odeurs du corps humain* (1885). At this period osphresiology exercised a new fascination.

6. See Chapter 11, p. 187.

7. Bordeu, *Recherches,* 435.

8. Yvonne Verdier, *Façons de dire, façons de faire* (Paris, 1979), especially pp. 20–77.

9. Bordeu, *Recherches,* 411.

10. Ibid., p. 414.

11. Quoted in ibid., p. 412.

12. Ibid., p. 413.

13. Brieude, "Mémoire," li.

14. Haller, *Eléments de physiologie,* 2:253. The theory underlies Monin's *Les Odeurs du corps humain.*

15. In Aristotle's opinion the humor became a cause of decay if it was not sufficiently "cooked by the heat of the body," and if the resultant products were not evacuated.

16. Bordeu, *Recherches,* 469.

17. In his notes to the 1844 edition of Cabanis's *Rapports du physique et du moral de l'homme,* L. Peisse stated with regard to specific bodily odors: "Among weak races or individuals, this odor is less marked; it is more strongly marked in very animalized species, in very vigorous bodies."

18. Ingenhousz, *Expériences sur les végétaux,* 151.

19. See the discussion in Chapter 13, p. 218.

20. Bordeu, *Recherches,* xlvii.

21. Ibid., p. 428.

22. For example, Abbé Armand-Pierre Jacquin, *De la santé, ouvrage utile à tout le monde* (1762), 283.

23. Xenophon, *Symposium;* Montaigne, "Des senteurs," in *Essais* (Paris, 1950), 351.

24. Chevalier Louis de Jaucourt, "Musc," in *Encyclopédie* (1765).

25. See Michèle Duchet, *Anthropologie et histoire au Siècle des Lumières* (Paris, 1977). In fact the author shows that there were not just one but several anthropological theories at that time (p. 409); the following discussion is in line with Duchet's analysis of Buffon's theory (pp. 199 ff.).

26. Brieude, "Mémoire," xlvii.

27. Duchet, *Anthropologie,* 203.

28. See Brieude, "Mémoire," lv, and Monin, *Les Odeurs du corps humain,* 51.

29. Brieude, "Mémoire," xlix.

30. See Virey, "Des odeurs," 249.

31. Jean-Noël Vuarnet makes this point in *Extases féminines* (Paris, 1980), 38–45. His book also contains a bibliography on the "odors of sanctity," often associated with *myroblitisme* and incorruptibility. On this subject see also Monin,

Les Odeurs du corps humain, 61. When alive, Saint Trévère smelled of roses, lilies, and incense, Saint Rose of Viterbo of roses, Saint Cajetan of oranges, Saint Catherine of violets, Saint Theresa de Avila of jasmine and irises, and Saint Lydwine of cinnamon bark (see Joris-Karl Huysmans, *Saint Lydwine de Schiedam* [Paris, 1901]). After death, Madeleine de Bazzi, Saint Stephen of Muret, Saint Philip Neri, Saint Paternien, Saint Omer of Thérouanne, and Saint François Olympias exhaled sweet odors. In the nineteenth century, alienists considered this phenomenon "the expression of a neurosis" (Monin, p. 61).

32. Brieude, "Mémoire," xlviii.
33. Landré-Beauvais, "Des signes tirés des odeurs," 423.
34. Cloquet, *Osphrésiologie,* 66.
35. Virey, "Des odeurs." This article gives the sources for travelers' numerous observations about the stench of savages.
36. Cloquet, *Osphrésiologie,* 15.
37. In the broad sense in which Buffon as well as Helvetius used the term, climate designated not only latitude and meteorological characteristics but also the nature of the soil and the inhabitants' way of life; that is, data on both the natural environment and the result of human adaptation (see Duchet, *Anthropologie,* 322).
38. Cloquet, *Osphrésiologie,* 66.
39. Brieude, "Mémoire," lx.
40. Ibid., p. l.
41. Ibid., pp. li–lii. The characters in Nougaret and Marchand's *Le Vidangeur sensible* compare the odors of the cesspool clearer and the butcher.
42. The topic also occurs in the Aristotelian *Problems.*
43. This progression reflected the breakdown in equilibrium that made putrefaction triumph within the living organism.
44. Bordeu, *Recherches,* 470.
45. Brieude, "Mémoire," lv.
46. Ibid., p. lxii, and Landré-Beauvais, "Des signes tirés des odeurs," 431.
47. H. A. P. A. Kirwan, *De l'odorat et de l'influence des odeurs sur l'économie animale* (1808), 26.
48. For an account of these experiments see Ingenhousz, *Expériences sur les végétaux,* 151 ff.
49. Louis Jurine, "Mémoire sur les avantages que la médecine peut retirer des eudiomètres," *Histoire et Mémoires de la Société Royale de Médecine,* 10 (1789), 19–100 (read August 28, 1787), describes minutely the method of removal used.
50. Jules-César Gattoni, ibid., p. 132.
51. At the most, Jurine's analysis of intestinal gases confirmed Claude Louis Berthollet's belief that wind originated in the putrid decomposition of meats.
52. Bordeu, *Recherches,* 523.
53. Quoted in Monin, *Les Odeurs du corps humain,* 239. The full significance of this observation emerges in the light of Bichat's definition of death.
54. On Hecquet see Ehrard, "Opinions médicales," 44; David Hartley, *Explication physique des sens, des idées et des mouvements tant volontaires qu'involontaires,* vol.

1 (Paris, 1755), 449–451; on the sympathists see Mauzi, *L'Idée du bonheur,* 313–314.

55. Charles-François Tiphaigne de la Roche, *L'Amour dévoilé ou le système des sympathistes* (1749), 45, 48, and 113.

56. Mirabeau, *Erotika Biblion* (1783), 19.

57. Preface to *Mémoires, 1744–1756* (Paris, 1977), lii.

58. This episode has been well studied by Gérard Wajeman, "Odor di femmina," *Ornicar,* 7 (1976), 108–110.

59. Dr. Augustin Galopin, *Le Parfum de la femme et le sens olfactif dans l'amour. Etude psycho-physiologique* (Paris, 1886). On this subject see p. 139.

60. Goethe, *Faust,* part II.

61. Verdier, *Façons de dire,* detects among the peasant women of Minot, in the Châtillonais, the persistence of beliefs in analogies among the cosmic, the mineral, and the human.

62. A.-A. Cadet de Vaux, "De l'atmosphère de la femme et de sa puissance," *Revue Encyclopédique,* 9 (1821), 445.

63. See Verdier, *Façons de dire,* 52 ff.

64. See Jean Borie, "Une Gynécologie passionnée," in *Misérable et glorieuse la femme du XIXᵉ siècle,* ed. J.-P. Aron (Paris, 1980), 152–189. On this subject see also Thérèse Moreau's works, for example, *Le Sang de l'histoire* (Paris, 1982).

65. Evariste-Désiré Desforges de Parny, "Le Cabinet de toilette"; François-Joachim de Pierres, cardinal de Bernis, "L'Eté," in *Les Saisons et les jours. Poèmes* (1764), sings of "the perfume of [nymphs'] blond tresses."

66. See Roland Barthes, *Fragments d'un discours amoreux* (Paris, 1977), 227.

67. Jean-Jacques Menuret de Chambaud, *Essai sur l'action de l'air dans les maladies contagieuses* (1781), 41.

68. Havelock Ellis, *Sexual Selection in Man,* vol. 4 of *Studies in the Psychology of Sex* (Philadelphia, 1905), 74.

69. At the end of the nineteenth century the theme was more amply developed. The psychologist Féré considered that this odor exercised a dynamic effect that could be used in industry; for example, weary ironing women got fresh energy from smelling the effluvia of their corsets; see Charles Féré, *Sensation et mouvement. Études expérimentales de psycho-mécanique* (1887), 50, and *La Pathologie des émotions* (1892), 440.

70. Rousseau, *Emile,* 201; Parny, "Le Cabinet de toilette."

71. As he confides in *L'Anti-Justine.*

72. As far as I know, vaginal odor was not mentioned in public and integrated into the range of odors that could be mentioned until Henry Miller's *Tropic of Capricorn.* The author considered initiation into these odors a rite of olfactory passage.

73. Jean-Baptiste Silva, "Dissertation où l'on examine la manière dont l'esprit séminal est porté a l'ovaire," in *Dissertations et consultations médicinales de MM. Chirac et Silva,* vol. 1 (1744), 188 ff., expanded this theme at length.

74. On this subject see Yvonne Kniebiehler's works.

75. *Putain* ("prostitute") from *puanteur* ("stink"). Restif opted for the popular etymology that derived *putain* from Latin *putida* ("stinking").

76. Silva, "Dissertation," 189. On the other hand, the ancients imagined that continence gave women a repulsive stench; Détienne, *Les Jardins d'Adonis*, 173. The interruption of conjugal relations, separation of sun and earth, caused the Lemnians' stink and the unpleasant, less powerful odor of the women of the Thesmophoria (ibid., p. 176). As far as I know, this aspect of the question was no longer mentioned in the eighteenth century.

77. Boissier de Sauvages, *Journal des Savants*, February 1746, p. 356; quoted in Fodéré, *Traité de médecine légale*, 6:232. The author was writing about the epizootic that prevailed in the Vivarais: when men "breathe the stinking blasts exhaled by the stomachs of these oxen, even when alive, from nearby, they are attacked by colics, followed by vomiting, and even diarrhea, which often makes the stomach swell up in an astonishing way."

78. Fodéré, *Traité de médecine légale*, 5:298.

79. "Man's breath is mortal for man," stated Rousseau; see François Dagognet, "La Cure d'air: Essai sur l'histoire d'une idée en thérapeutique médicale," *Thalès*, 10 (1959), 87.

80. Verdier, *Façons de dire*, found this at Minot. An old woman reported of a friend: "She caught it like an epidemic to my sister with her breath" (p. 46).

81. Etienne Senancour, "Les Introuvables," in *Obermann*, 2 vols. (1844), 2:48.

82. Arbuthnot, *Essay Concerning Effects of Air*, 189.

83. Boissier de Sauvages, *Dissertation*, 56.

84. Charles-Polydore Forget, *Médecine navale ou nouveaux éléments d'hygiène, de pathologie et de thérapeutique médico-chirugicales*, 2 vols. (1832), 1:332. Note that this text is later than the other evidence quoted in this chapter.

85. Stephen Hales, *A Description of Ventilators* (London, 1743), 51.

86. Forget, *Médecine navale*, 1:184. The following paragraphs are a synthesis of numerous contemporary descriptions, particularly those by Duhamel-Dumonceau, *Moyens de conserver la santé*.

87. Forget, *Médecine navale*, 1:29.

88. Ibid., p. 186.

89. Ibid. See also Fodéré, *Traité de médecine légale*, 6:476 ff.

90. Alexandre Parent-Duchâtelet, *Recherches pour découvrir la cause et la nature d'accidents très graves, développés en mer, à bord d'un bâtiment chargé de poudrette* (1821). The rest of the crew was ill.

91. "The whirlpool of their own transpiration" could not "get lost in the air," wrote Duhamel-Dumonceau, *Moyens de conserver la santé*, 30.

92. Hales, *Description of Ventilators*, 38–39.

93. For example, the vicomte de Morogues, calculating the weight of vapors transpired or exhaled in a thirty-cannon frigate, concluded that the volume of foul exhalations equaled almost five cubic feet of water; Duhamel-Dumonceau, *Moyens de conserver la santé*, 44.

94. Macquer, Lavoisier, Fourcroy, and Vicq d'Azyr were foremost among its members.

95. Mercier, *Tableau de Paris*, 8:1.

96. Howard, *State of Prisons*, 214.

97. Casanova, *Mémoires*, 547 and 588.

98. Senancour, *Obermann*, 1:83.

99. Libretto of *Fidelio* by Beethoven, translated and adapted from J.-N. Bouilly; the prisoners are authorized by Rocco to take the air for a second.
100. Jules Michelet, *Histoire de France*, 17 vols. (1833–67), 13:317–318.
101. Howard, *State of Prisons*, 6–7.
102. Bacon, *Sylva Sylvarum*, quoted in Pringle, *Observations*, 345.
103. Howard, *State of Prisons*, 9.
104. Ibid.
105. Pringle, *Observations*, 347–348.
106. Fodéré, *Traité de médecine légale*, 5:311.
107. The incident is cited over and over again. See, for example, Dr. Jean-Baptiste Banau and François Turben, *Mémoire sur les épidémies du Languedoc* (1786), 12.
108. See Chapter 2, p. 24.
109. For this period it is anachronistic to distinguish between prison and hospital; at the end of the eighteenth century, however, the distinction was beginning to be somewhat justified.
110. That "horrible mixture of putrescent infection" was the cemetery of the poor, wrote Léopold de Genneté: "breath is infected, wounds putrefy, sweat smells of corpses"; *Purification de l'air croupissant dans les hôpitaux, les prisons, et les vaisseaux de mer . . .* (1767), 10.
111. It is noteworthy that references to smells become fewer when he is dealing with English establishments.
112. Mercier, *Tableau de Paris*, 8:7 and 8.
113. Jacques-René Tenon, *Mémoires sur les hôpitaux de Paris* (1788).
114. Michel Foucault, in M. Foucault, F. Beguin, B. Fortier, A. Thalamsy, and B. Barret-Kriegel, *Les Machines à guérir, aux origines de l'hôpital moderne* (Paris, 1979).
115. Tenon, *Mémoires sur les hôpitaux*, 208.
116. Ibid., p. 223.
117. Ibid., p. 238.
118. Even before barracks became very numerous, the general staff attributed the great epidemics that ravaged the French army in 1743 to overcrowding and the stagnation of the air; André Corvisier, *L'Armée française du XVIIe siècle au ministère de Choiseul. Le soldat*, vol. 2 (Paris, 1964), 672.
119. Mercier, *Tableau de Paris*, 7:309. He condemned balls for this reason.
120. Senancour, *Obermann*, 2:48.
121. Ibid., p. 191.
122. Edna Hindie Lemay, "La Vie parisienne des députés de 89," *L'Histoire*, 44 (1982), 104.
123. Favre, *La Mort dans la littérature*, 252, and, on Voltaire's involvement in the debate, 259.
124. Ariès, *L'Homme devant la mort*, 474–475.
125. He even nursed a plan to write a history of odors, collecting "everything scattered in the works of writers on this subject"; *De morbis artificum*, 199.
126. Ibid., p. 513.
127. Ibid., p. 336. The fear of emanations from laundries persisted for a long time; under the July Monarchy, when the new requirements of cleanliness were spreading in Paris, the vapors of washing gave rise to numerous complaints to the Conseil de Salubrité.

128. Ibid., pp. 152–153.
129. With the exception of the Jews, however (see Chapter 9, p. 145); but the way in which this belief took root in the religious history of the West is familiar.
130. Mercier, *Tableau de Paris*, 1:137–138. On the same subject, see 126–130.
131. Françoise Boudon, "La Salubrité du grenier de l'abondance à la fin du siècle," *XVIII< Siècle*, 9 (1977), 171–180.
132. See Fortier, "La Politique de l'espace parisien."
133. Boudon, "La Salubrité," 176.
134. Jurine measured samples of "the air from the beds"; "Mémoire," 71 ff. For the scale of unhealthy air in "inhabited apartments," see 90–91.

4. Redefining the Intolerable

1. Antoine Tournon, *Moyen de rendre parfaitement propres les rues de Paris* (1789), 60.
2. Daniel Roche, *Le Siècle des Lumières en province: Académies et académiciens provinciaux*, vol. 1 (Paris, 1978), 378.
3. Mercier, *Tableau de Paris*, 1:222.
4. Chauvet, *Essai sur la propreté*, 18.
5. Mercier, *Tableau de Paris*, 1:267.
6. Favre, *La Mort dans la littérature*, 40.
7. See Madeleine Foisil, "Les Attitudes devant la mort au XVIII< siècle: Sépultures et suppressions de sépultures dans le cimetière parisien des Saints-Innocents," *Revue Historique*, April–June 1974, p. 322.
8. Fortier, "La Politique de l'espace parisien," 34.
9. Young, *Travels in France*, 39.
10. Ibid., p. 33.
11. Ibid., pp. 160–161.
12. See Jean Delumeau, *La Peur en Occident* (Paris, 1978), 129 ff.
13. Menuret, *Essai sur l'action de l'air*, 51.
14. Foisil, "Les Attitudes devant la mort," 311.
15. Cadet de Vaux, *Mémoire historique*, drew up the chronology of these complaints.
16. Mercier, *Tableau de Paris*, 8:340; the italics are mine.
17. Ibid., p. 341.
18. The ordinance of 1726 had already prohibited the first from injuring the second.
19. Laborie, Cadet the younger, and Parmentier, *Observations sur les fosses*, 105.
20. Géraud, *Essai sur la suppression des fosses*, 43.
21. This indignation, though, persisted for more than half a century without any result.
22. Thouret, *Rapport sur la voirie*, 4.
23. Ronesse, *Vues sur la propreté*, 28.
24. Edmond Huot de Goncourt and Jules Huot de Goncourt, *La Femme au XVIII< siècle* (1862), 368.
25. Louis Damours, *Mémoire sur la nécessité et les moyens d'éloigner du milieu de Paris, les tueries de bestiaux et les fonderies de suif* (1787), 9.
26. Boudon, "La Salubrité," 172. Tournon's evidence confirms this point.

27. Géraud, *Essai sur la suppression des fosses,* 49 and 41.
28. Emphasized particularly by Maurice Agulhon in the political field.
29. Guillerme, "Le Malsain," 65.
30. Pierre Chaunu, quoted in Foisil, "Les Attitudes devant la mort," 323. It is impossible to measure the change objectively; the historian is totally dependent on necessarily subjective witnesses.
31. See Fortier, "La Politique de l'espace parisien," 19, referring to an article in the *Journal de Paris,* July 25, 1781.
32. Laporte, *Histoire de la merde,* 60.
33. Ibid., p. 18.
34. Marcel Mauss, *Sociologie et anthropologie* (Paris, 1980), emphasizes the role of Kant and even more of Fichte.
35. Fortier, "La Politique de l'espace parisien," 41.
36. Menuret, *Essai sur l'action de l'air,* 51.
37. In 1835 Mrs. Trollope (*Paris and the Parisians in 1835,* 2 vols. [1836]), choked by the stench of the Continent, tried to understand the sensory revolution taking place; it seemed to her, probably quite rightly, that its progress was faster in England. Her premonitory analysis partly supported the theories formulated earlier. The "improvement in English delicacy has been gradual, and in very just proportion to the increase of her wealth, and the fastidious keeping out of sight of everything that can in any way annoy the senses" (1:229–230). "This withdrawing from the perception of the senses everything that can annoy them—this lulling of the spirit by the absence of whatever might awaken it to a sensation of pain,—is probably the last point to which the ingenuity of man can reach in its efforts to embellish existence" (1:233). But perhaps this excessive refinement would hurl England into the gulf where civilizations perish. The British lady therefore minimized the medical goal, which would necessitate allusion to bodily functions, and stressed the desire for delicacy, the supreme but dangerous caress of the soul. In her opinion, the purification of the language did not precede the purification of space but resulted from it.
38. In the eighteenth century, "odors" designated perfumes.
39. Pierre-Joseph Buc'hoz, *Toilette de flore a l'usage des dames* (1771), 192.
40. "Des odeurs," cited earlier, and "De l'osmologie, ou histoire naturelle des odeurs," *Bulletin de Pharmacie,* 4 (May 1812), 193–228.
41. "Observations sur les parties volatiles et odorantes des médicaments tirés des substances végétales et animales: Extraites d'un mémoire de feu M. Lorry, par M. Hallé," *Histoire et Mémoires de la Société Royale de Médecine,* 7 (1784–85), 306–318.
42. Nicolas Lémery, *Pharmacopée universelle,* 892.
43. Banau and Turben, *Mémoire,* 90.
44. Ramazzini, *De morbis artificum,* 198.
45. Lémery, *Pharmacopée universelle,* 896 and 914.
46. A very old belief was involved here; see Delumeau, *La Peur en Occident,* 114.
47. Nicolas Blégny, *Secrets concernant la beauté et la santé . . . recueillis par M. Daquin,* 2 vols. (Paris, 1688–89), especially 2:696.
48. See Françoise Hildesheimer, "La Protection sanitaire des côtes françaises au

XVIIIᵉ siècle," *Revue d'Histoire Moderne et Contemporaire,* 27 (July–September 1980), 443–467. In the Levant, health procedures combined open air, "perfume," and isolation.

49. See E. H. Ackerknecht, "Anticontagionism between 1821 and 1867," *Bulletin of the History of Medicine,* 1948, pp. 562–593.
50. Pringle, "Experiments," 528–534.
51. Lind, *Essay on the Health of Seamen,* 102.
52. Boissieu, *Dissertations sur les antiseptiques,* 67.
53. Bordenave, ibid., pp. 190 ff.
54. The correlation established between the balsamic and the salubrious, the malodorous and the insalubrious, always remained fragile. Scientific theories lost their coherence in practice. They fitted in with one another, amalgamated, or overlapped like the tiles of a roof. Becher was convinced of the beneficial properties of foul-smelling excrement, and, well before Ingenhousz delivered a valid analysis of photosynthesis, doctors had denounced the noxious effect of certain balsamic plants.
55. Goncourt and Goncourt recalled that during the Regency people called the service celebrated before dinner in the chapel of Saint-Esprit "musked mass"; *La Femme au XVIIIᵉ siècle,* 395.
56. Père Jean-Pierre Papon, *De la peste ou époques mémorables de ce fléau et les moyens de s'en préserver,* 2 vols. (Paris, year VIII), 2:47.
57. Buc'hoz, *Toilette de flore,* 7.
58. Lémery, *Pharmacopée universelle,* 892. He advised particularly that a balm composed of musk, ambergris, civet, and storax that "resisted bad air by its strong odor" be provided "so that one might be able to smell it often."
59. According to Baumes's evidence, *Mémoire,* 224. The traveler who had crossed a swampy region, arriving at the hostelry in the evening, would burn sulfur in his room, imbibe infusions of odoriferous herbs, smoke tobacco "or any other aromatic substance," and make every effort not to swallow his saliva (p. 226).
60. Louis-Bernard Guyton de Morveau, *Traité des moyens de désinfecter l'air* (1801), 149; however, he expressed doubt about the efficacy of the practice.
61. Baumes, *Mémoire,* 224.
62. Ramazzini, *De morbis artificum,* 209.
63. Commentary on ibid. in *Essai sur les maladies des artisans traduit du latin de Ramazzini avec les notes et des additions par M. de Fourcroy* (1777), 332.
64. Alexandre Parent-Duchâtelet, *Rapport sur le curage des égouts Amelot, de la Roquette, Saint-Martin et autres,* reprinted in *Hygiène publique,* 1:364; the workers first used sachets and then bottles that emitted odors of chlorine.
65. Duhamel-Dumonceau, *Moyens de conserver la santé,* 132 ff.
66. According to Joseph-Marie-François Delassone senior and Claude-Melchior Cornette, "Mémoire sur les altérations que l'air éprouve par les différentes substances que l'on emploie en fumigation . . . ," *Histoire et Mémoires de la Société Royale de Médecine,* 8 (1786), 324. Information on these methods can also be found in Hales, *Description of Ventilators,* 46, and Hallé, "Air," in *Encyclopédie méthodique. Médecine* (1787), 572–575.
67. *Encyclopédie,* s.v. "Parfumoir."

68. In 1796 Jackson and Moser offered Londoners their fumigating lamp, intended especially for the combustion of the chemical products that composed the new range of disinfectants; Guyton de Morveau, *Traité*, 147.
69. The process is described by Ramel, *De l'influence des marais*, 301.
70. Papon, *De la peste*, 1:329.
71. See especially Fodéré, *Traité de médecine légale*, 6:159.
72. This is Tenon's opinion in *Mémoires sur les hôpitaux*, 451. He claims to have known the elder Lind well.
73. Quoted in Duhamel-Dumonceau, *Moyens de conserver la santé*, 138.
74. Jean Antoine Chaptal, *Eléments de chimie*, vol. 3 (1803), 111.
75. Felix Vicq d'Azyr, *Instruction sur la manière de désinfecter une paroisse* (1775), 7–8.
76. "The immense storerooms full of this plant and other drugs were a powerful obstacle to the invasion"; Menuret, *Essai sur l'action de l'air*, 60.
77. *Histoire et Mémoires de la Société Royale de Médecine*, 3 (1782), 44, quoted in Baumes, *Mémoire*, 164. Which justified the hymn to smoke: "the good quality of the air in the big cities is partly due to its services" (p. 163).
78. "It is important that limekilns, glassworks, soapworks, brandy or oil of vitriol distilleries be erected in these unhealthy districts, according to area and circumstances; these establishments would be doubly useful, since they would serve to correct the air and provide the inhabitants . . . with work"; Baumes, *Mémoire*, 165. More optimistic than a number of his contemporaries, Baumes added: "Moreover, the combustion of coal in simply constructed ovens would combine the advantage of avoiding wood consumption . . . with the advantage of spreading with a great deal of smoke sulfurous emanations, which have an undisputed purifying property."
79. See Blégny, *Secrets*, 2:167, "Parfum pour la guérison de la vérole."
80. For examples see Jaucourt, "Musc."
81. Quoted in ibid.
82. Virey, "Des odeurs," 174, and Hartley, *Explication physique des sens*, 331.
83. See de Sèze, *Recherches physiologiques*, 159. The author does not share Buffon's views.
84. Virey, "Des odeurs," 254.
85. See Paul Dorveaux, *Histoire de l'eau de la Reine de Hongrie* (Paris, 1921), 6. Blégny, *Secrets*, 1:684, listed the many virtues of Eau de la Reine de Hongrie and pointed out that "many people like its strong odor and sniff it incessantly."
86. Lorry, "Observations sur les parties volatiles," 318.
87. Fourcroy, *Essay de Ramazzini*, 128.
88. Ibid., p. 221.
89. Virey, "De l'osmologie," 206. Fiery temperaments in particular should refrain from voluptuous smells, which Virey thought caused them to exhale fetid odors, thus paralleling the issue of semen. Foul odors, when powerful, could exert the same noxious effects. "I have several times observed," Ramazzini remarked, "women who live near these [candlemakers'] stalls complain of hysterical passions because of the bad odor"; *De morbis artificum*, 180–181. For that reason, he warned men of letters against overwork at night. In *De morbis ab immunditiis* Platner, a Leipzig doctor, cataloged the risks incurred from breathing disgusting odors.

90. For examples see Boissier de Sauvages, *Dissertation,* 56, and above all Cloquet, *Osphrésiologie,* 80–98, who based himself particularly on Thomas Cappelini, *Mémoire sur l'influence des odeurs,* as well as on Triller's observations. Tobacco itself, Cloquet stated (p. 352), concealed deadly effects. Heavy smokers ceased to be able to smell; the drug slowly destroyed their olfactory nerves, as was proved by dissections of smokers' heads.
91. Mercier, *Tableau de Paris,* 6:47: "A guard was kept seated near the door and sniffed everyone who arrived. She repeated incessantly: 'Do you have any odors?' "
92. Howard, *Principal Lazarettos,* 59.
93. Virey, "De l'osmologie," 216.
94. This was a very old quarrel. Plato in the *Republic* heaped anathema on perfumes that encouraged indolence and pleasures. In ancient Greece the use of aromatics was a characteristic of the courtesan, "seductive illusion of a life all in perfumes." The more limited the part played by perfume in sexual intercourse, the greater its legitimacy seemed (Jean-Pierre Vernant, Preface to Détienne, *Les Jardins d'Adonis,* xiii and xxxvi). It is true that the animal substances used in eighteenth-century France were not known in antiquity; far from constituting a putrid threat, the ancient aromatic, associated with ideas of heat and dryness and formed close to the celestial fire, symbolized the imputrescible. For the denunciation of luxury see Abbé François André Adrien Pluquet, *Traité philosophique et politique sur le luxe,* 2 vols. (1786).
95. Jacquin, *De la santé,* 290 ff.
96. Luigi Antonio Caraccioli, *La Jouissance de soi-même* (1759), 333.
97. See Georges Vigarello, *Le Corps redressé* (Paris, 1978), 87 ff.
98. See Thorstein Veblen, *The Theory of the Leisure Class* (London, 1924), 154 (1st ed. 1899).
99. De Genneté, *Purification,* 11.
100. Vicq d'Azyr, *Instruction,* 8.
101. Jacquin, *De la santé,* 82.
102. Guyton de Morveau, *Traité,* 93.
103. The numerous experiments conducted by Delassone senior and Cornette ("Mémoire") proved beyond doubt that fumigation with aromatic substances did nothing but mephitize the bell jar. The researchers did not know that this phenomenon resulted from combustion and that it was not sufficient grounds for challenging the therapeutic value of the substances they analyzed.
104. Quoted in Guyton de Morveau, *Traité,* 138 and 139.
105. See Fourcroy, "Air atmosphérique," in *Encyclopédie méthodique. Médecine* (1787), 577.
106. Jean-Noël Hallé et al., *Codex des médicaments ou pharmacopée française* (1818).

5. The New Calculus of Olfactory Pleasure

1. For example, Jacquin, *De la santé,* 283: "Cleanliness is a concern to avoid everything that can revolt the delicacy of the senses; it is one of the principal virtues of society." On the evolution of Lasallean courtesy in the schools, see Roger Chartier, Dominique Julia, and Marie-Madeleine Compère, *L'Education en France du XVIᵉ au XVIIIᵉ siècle* (Paris, 1976), 143–144.

2. Platner's ideas are expounded in Baumes, *Mémoire,* 189. An analogous theory is expounded in Moheau (Antoine Auget, baron of Montyon), *Recherches et considérations sur la population de la France* (1778), 109.

3. Quoted in Baumes, *Mémoire,* 191.

4. Hallé, *Recherches sur la nature du méphitisme,* 111.

5. See Lion Murard and Patrick Zylberman, "Sanitas sanitatum. Equipement sanitaire et hygiène sociale" (Centre d'Etudes de Recherches et de Formation Institutionelles, 1981), 275–280; see also Jean-Maurice Bizière, "Before and After: Essai de psycho-histoire," *Revue d'Histoire Moderne et Contemporaine,* 27 (April–June 1980), 177–207.

6. Daniel Roche's *Le Peuple de Paris* (Paris, 1981) makes some qualification necessary; inventories taken after death revealed a not inconsiderable number of flagons and basins in certain plebeian interiors, notably within groups in contact with the aristocracy (p. 122).

7. See Bruno Fortier, "La Maîtrise de l'eau," *XVIIIᵉ siècle,* 9 (1977), 193–201.

8. See especially Ronesse, *Vues sur la propreté,* 91. He wrote in 1782: "The water that comes from houses is infinitely more abundant than it was fifteen years ago: this results from the very frequent use of baths, which doctors are today prescribing for many more diseases than they did in former times, and from the taste that the public has acquired for this practice; with the result that there are baths in all newly built houses and that when an individual *in comfortable circumstances* wants to rent an apartment he regards a bathroom as one of the most essential rooms" (my italics). The custom grew up of fitting pumps to almost every well. As a result, domestic servants in big houses stopped being economical with water; they used it for washing courtyards, kitchens, and even carriages.

9. *Recherches et considérations,* 110.

10. On the establishment of discipline at school see Chartier, Julia, and Compère, *L'Education,* especially 145.

11. M. Déjean (Antoine de Hornot), *Traité des odeurs* (1764), 147.

12. Chartier, Julia, and Compère, *L'Education,* 144.

13. Mauzi, *L'Idee du bonheur,* 427.

14. Déjean, *Traité des odeurs,* 457.

15. S.v. "Parfum." The same opinion is expressed in Jaucourt, "Musc."

16. Le Cat, *Traité des sensations,* 256. Musk gave vapors, fainting fits, to all ladies and some men.

17. Déjean, *Traité des odeurs,* 91.

18. Indeed, the use of musk persists today, although its function has changed; it now tends to be reserved for men. Animal perfume has become a symbol of virility but it has lost all reference to rut.

19. Ellis, *Sexual Selection in Men,* 99 and 91. When Ellis was writing, at the beginning of the twentieth century, experts had long emphasized the impact of that odor on sexual behavior. In the early nineteenth century Esquirol reported several cases of women driven mad by inhaling musk during lactation. Around 1880 Féré confirmed that musk was more reminiscent of sexual secretions than was any other perfume.

20. Iwan Block Hagen, *Die Sexuelle Osphrésiologie* (Leipzig, 1901), and Ellis, *Sexual Selection in Man,* contain excellent bibliographies on the subject.

21. Ellis, *Sexual Selection in Man,* 99.
22. Freud wrote: "The fateful process of civilization would thus have set in with man's adoption of an erect posture. From that point the chain of events would have proceeded through the devaluation of olfactory stimuli and the isolation of the menstrual period to the time when visual stimuli were paramount and the genitals became visible, and thence to the continuity of sexual excitation, the founding of the family and so to the threshold of human civilization"; "Civilization and Its Discontents," in *The Standard Edition of the Complete Psychological Works of Sigmund Freud,* ed. James Strachey, vol. 21 (London, 1975), 99–100, no. 1.
23. Hartley, *Explication physique des sens,* 332.
24. Déjean, *Traité des odeurs,* 8 ff. This trend must be related to the development that affected the spectrum of colors in clothing fashions; soft tints prevailed at the same time that jonquil triumphed over musk. See Roche, *Le Peuple de Paris,* 177.
25. Already used a great deal in the seventeenth century; see Blégny, *Secrets,* 1:687.
26. Malouin, *Chimie médicinale,* 275.
27. See Dr. Louis Reutter de Rosemont, *Histoire de la pharmacie à travers les âges,* vol. 2 (Paris, 1931), 438.
28. Ibid., p. 441. As early as 1740 the Peruvian heliotrope was introduced to France by Joseph de Jussieu.
29. Casanova, *Mémoires,* 255.
30. This was notably Fourcroy's opinion, *Essai de Ramazzini,* 186.
31. See Blégny, *Secrets,* 1:697, and Déjean, *Traité des odeurs,* 303.
32. Restif de La Bretonne, *L'Anti-Justine,* passim.
33. For example, he washed his Venetian nun's "superb neck with rose water"; *Mémoires,* 448.
34. As in the novel by Charles Rochette de La Morlière, *Angola, histoire indienne* (1746) (2 vols. in 1).
35. Roland Barthes, *Sade, Fourier, Loyola* (Paris, 1971). Sade's scenario for the body involved sight; flowers and excrement figured in it only to mark out the path to degradation. "Written down, excrement does not smell; Sade can drench his opponents in it, we receive no effluvia from it, only the abstract sign of something disagreeable" (p. 140). Sade's writing does, however, contain some references to breath, to the odor of sperm, and, of course, to sulfur (as in la Durand's sorcery in *Juliette*).
36. Déjean, *Traité des odeurs,* 423.
37. Ibid., p. 431.
38. Buc'hoz, *Toilette de flore,* part 1. See also Mercier, *Tableau de Paris,* 6:153.
39. Mme. Campan, *Mémoires sur la vie de Marie-Antoinette, reine de France et de Navarre* (1849), 97, notes the proliferation of flowers in hair styles at the court of Louis XVI.
40. Mercier, *Tableau de Paris,* 2:158.
41. Casanova, *Mémoires,* 295. He has the same type of reaction on p. 176.
42. Ibid., p. 185.
43. Ibid., p. 139.
44. Alexandre Dumas, "Les Parfums," *Le (Petit) Moniteur Universel du Soir,* October 16, 1868.

45. Treatise on perfumes, *Le Parfumeur royal* (1761), 83.
46. Déjean, *Traité des odeurs*, 4.
47. Casanova, *Mémoires*, 427.
48. For example, letters XIV (June 1783) and XXIII (March 8, 1784), *Lettres choisies du marquis de Sade* (Paris, 1963), 169 and 222.
49. Casanova, *Mémoires*, 435.
50. *Le Parfumeur royal*, 150.
51. Déjean, *Traité des odeurs*, 447.
52. There is considerable evidence on this subject: *Encyclopédie*, s.v. "Parfum"; Buc'hoz, *Toilette de flore*, 137; *Le Parfumeur royal*, 7.
53. *Le Parfumeur royal*, 152–153.
54. Ibid., p. 158.
55. Buc'hoz, *Toilette de flore*, 67.
56. Ibid., p. 233.
57. *Le Parfumeur royal*, 158, 159, 148–149, 202.
58. Evariste-Desiré Desforges de Parny, "Le Cabinet de toilette," in *Oeuvres de Parny, élégies et poésies diverses*, 3 vols. (1861), 3:78–79; Rousseau, *Emile*, 201. Godard d'Aucourt, *Thémidore* (Paris, 1980), 226 (1st ed. 1745), suggested that the use of subtle and delicately allusive perfumes was a favorite weapon of the "devout" female libertine. The delicate scents of Mme. de Dorigny's toilette prepared the way for Thémidore's downfall.
59. René-Louis Girardin, *De la composition des paysages* (1777), 59.
60. Senancour, *Obermann*, 1:71.
61. Ibid.
62. Louis-François Ramond de Carbonnières, *Observations faites dans les Pyrénées pour servir de suite à des observations sur les Alpes* (1789), 346.
63. Favre, *La Mort dans la littérature*, 251.
64. Jurine, "Mémoire," 95.
65. Dagognet, "La Cure d'air," gives a very subtle analysis of this "social hypnotherapy," this "emotional hibernation" (p. 85), this "delirium formed around the vivifying air of the mountains" (p. 76). To explain this desire to tap the energy of the peaks, he refers to Jung's themes of aerial resurrection and salvific aspiration. The fashion for staying in the mountains must be linked to the more general fashion for "ascensional" ventures and fantasies.
66. See, for example, Tourtelle, *Eléments d'hygiène*, 271.
67. Géraud, *Essai sur la suppression des fosses*, 95.
68. Horace Benedict de Saussure, *Voyages dans les Alpes*, 4 vols. (Neuchâtel, 1779–96), 1:518: "The air in mountains rising over five or six hundred fathoms above sea level, is vitiated by other exhalations."
69. Senancour, *Obermann*, 1:54; Ramond, *Observations dans les Pyrénées*, 348.
70. Senancour, *Obermann*, 2:174: Saussure, *Voyages dans les Alpes*, 2:480 ff.
71. Girardin, *De la composition des paysages*, 128.
72. Claude-Henri Watelet, *Essai sur les jardins* (1764), 34.
73. This is a chapter title in Girardin, *De la composition des paysages*.
74. Mercier, *Tableau de Paris*, 10:72: "Whoever does not like the odor of new-mown hay does not know the pleasantest of perfumes"; Ramond, *Observations dans les Pyrénées*, 88; Loaisel de Tréogate, *Dolbreuse* (1783), 81.
75. Senancour, *Obermann*, 1:23. Beatrice Le Gall, *L'Imaginaire chez Senancour*, 2

vols. (Paris, 1966), 1:43, studies this sensation. The odor of mown hay is one of the symbols of early adolescence.

76. Liane Lefaivre and Alexander Tzonis, "La Géométrie du sentiment et le paysage thérapeutique," *XVIIIe Siècle*, 9 (1977), 74.

77. Girardin, *De la composition des paysages*, 123.

78. C. C. L. Hirschfeld, *Théorie de l'art des jardins*, 5 vols. (Leipzig, 1779–85), 1:185 and 186.

79. Thomas Whately, *Observations on Modern Gardening* (London, 1771); Jean-Marie Morel, *Théorie des jardins* (1776).

80. Hirschfeld, *Théorie de l'art des jardins*, 1:185.

81. Girardin, *De la composition des paysages*, 52.

82. Horace Walpole, *Essay on Modern Gardening* (London, 1785). Hirschfeld, *Théorie de l'art des jardins*, 2:94, writes: "It is in places where man rests, where he gives himself up to his thoughts and his imagination, where he prefers feeling to reflection, that the families of odoriferous flowers should spread their sweet, balsamic, refreshing perfumes and thus enhance the sensation of the delights of creation by satisfying a new sense. Let places intended for rest and sleep, let studies, dining rooms, baths be surrounded by sweet odors of violet, lily, stock . . . cloves . . . white narcissus, white lily, hyacinth, carnation, mignonette or Egyptian reseda . . . jonquil, etc. Enjoyment of these perfumes in an indescribable way spreads a sort of re-creation and calm within man and pours peace and a feeling of well-being into his soul, which gently warms him."

83. John Milton, *Paradise Lost*, book 5, 294. Books 4 and 5 of *Paradise Lost* praise the natural perfumes of flowers and meadows. Milton in his blindness appealed first to his reader to imagine scents: fragrant bushes, roses, jasmine, and violets perfume the bower and, more accurately, the secret retreat that shelters the loves of Adam and Eve.

84. Girardin, *De la composition des paysages*, 48.

85. Ibid., p. 132.

86. Hirschfeld, *Théorie de l'art des jardins*, 1:51.

87. Watelet, *Essai sur les jardins*, 34.

88. Ellis, *Sexual Selection in Man*, 102. Even the chastest of women, when smelling a flower deeply, closed her eyes and, "if very sensitive, trembles all over, presenting an intimate picture which otherwise she never shows, except perhaps to her lover." Ellis noted that some nineteenth-century moralists had condemned flowers solely because of this effect.

89. Loaisel de Tréogate, *Dolbreuse*, 174 and 80.

90. Andréa de Nerciat, *Félicia ou mes fredaines* (Paris, 1979), 196 (1st ed. 1776). For another example see Rochette de la Morlière, *Angola*, 2:16.

91. Hirschfeld, *Théorie de l'art des jardins*, 5:66.

92. Ibid., p. 19.

93. Ramond, *Observations dans les Pyrénées*, 165.

94. See Jean Starobinski, *La Transparence et l'obstacle* (Paris, 1971), 196.

95. Ramond, *Observations dans les Pyrénées*, 88. This example has often been quoted: Maine de Biran mentions it in his *Journal* (vol. 1 [Neuchâtel, 1954], 151). Mountainsides were particularly favorable for the manifestation of the memorative sign; because of their calm, their silence, and the fatherly proximity

of the sun, they summoned up the image of the mother and were therefore propitious to the reliving of childhood. All these themes were expanded by Michelet (see Dagognet, "La Cure d'air," 81 ff.).

96. Senancour, *Obermann,* 2:58.

97. Castan, communication to a colloquium on the history of prisons, held at l'Ecole des Hautes Etudes en Sciences Sociales December 19, 1980; Mandrou, *Introduction à la France moderne,* 70 ff.

98. Jean-François Saint-Lambert, *Les Saisons* (1769), 35, quoted in Mauzi, *L'Idée du bonheur,* 320.

99. Senancour, *Obermann,* 2:268.

100. Mauzi, *L'Idée du bonheur,* 114.

101. Rousseau felt the botanist's attraction for flowers. He was more interested in admiring their arrangement than in smelling them, more in "resting from ecstasy than nourishing it"; Le Gall, *L'Imaginaire,* 1:331. The herbarium he made himself was primarily a "reminder"; it was through sight that he expected to conjure up the immediate presence of memory; Starobinski, *La Transparence et l'obstacle,* 197.

102. See p. 140.

103. Caraccioli, *La Jouissance,* quoted in Mauzi, *L'Idée du bonheur,* 195.

104. Franklin, *La Vie privée,* 7:31.

105. Fodéré, *Traité de médecine légale,* 6:526.

106. Senancour, *Obermann,* 2:269 and 268.

107. Mauzi, *L'Idée du bonheur,* 317.

108. Quoted in ibid., p. 319.

109. Senancour, *Obermann,* 1:113.

110. Ibid., pp. 244–245.

111. Marcel Raymond, *Senancour, sensations et révélations* (Paris, 1965), analyzed this search for happiness through sensation in Senancour's work. He saw in the special sensitivity to odor the true appreciation of the first stirring of the heart and compared this feeling with the sense of emergence in Novalis. Le Gall, *L'Imaginaire,* demonstrated that violets and jonquils conjured up for Senancour two experiences of love. She added (1:271): "He loves violets because sometimes when they are hidden beneath the grass they are nothing but a fragrance." In fact Senancour wrote in his *Rêveries* (Paris, 1939), 1:100–101: "The feeling that emanates from them is offered to us and immediately refused, we seek it in vain, a light breeze has carried its perfume away, it brings it back and carries it away again, and its invisible caprice has created our pleasure." Senancour was, like Friedrich Hoffmann *(Der goldener Topf),* fascinated by the search for sensory correspondences and inspired by the research of Rev. Père Castel; he too dreamed of a keyboard of odors well before Huysmans's Des Esseintes (Le Gall, *L'Imaginaire,* 1:331).

112. Note, though, that *Obermann* did not appear until 1803.

6. The Tactics of Deodorization

1. Jean-Claude Perrot, *Genèse d'une ville moderne. Caen au XVIIIᵉ siècle,* 2 vols. (Paris, 1975), 1:9; 2:945, 950; and 1:10.

2. See Gilles Lapouge, "Utopie et hygiène," *Cadmos,* 9 (1980), 120.

3. This was the opinion of Dr. Lecadre, "Le Havre considéré sous le rapport hygiénique," *Annales d'Hygiène Publique et de Médecine Légale*, 42 (1849), 255.
4. See Ramel, *De l'influence des marais*, 251.
5. Bertholon, *De la salubrité*, 69. Anxiety is the predominant tone of all writers on the subject.
6. Boudon, "La Salubrité," 178.
7. The paving alongside the convent of the Jacobins was replaced four times in forty years; J.-C. Perrot, *Genèse d'une ville moderne*, 95.
8. For example, Baumes, *Mémoire*, 179. Howard, *State of Prisons*, 23.
9. For example, the police ordinance of November 8, 1729.
10. Cited in Chevreul, "Mémoire," 30.
11. Fodéré, *Traité de médecine légale*, 6:256.
12. The wish to be fortified against the threat of infection was connected with the same desire. Scientists imaginatively suggested the use of some complicated machines. Fourcroy, for example, recommended that starchworkers "put a sort of paper funnel round their necks, with its widest side toward their heads, in order to divert the course of the vapor that would otherwise strike their faces"; *Essai sur Ramazzini*, 313. Pharmacists suggested curious antimephitic glazes. Banau and Turben (*Mémoire*, 99) prepared for this purpose a concoction that could be used to smear frock coats. This was no aberrant case; Fodéré (*Traité de médecine légale*, 6:112) recommended that his colleagues, patients' families, and neighbors also use glazed taffeta wrappers to cover clothes, boots, and hats.
13. Howard, *State of Prisons*, 342. Roche, *Le Peuple de Paris*, 140, notes the increase in tapestry hangings in the houses of the masses; at the end of the century they embellished 84 percent of interiors.
14. See Chapter 7, pp. 123–124.
15. Bertholon, *De la salubrité*, 97; Thouret, *Rapport sur la voirie*, 10.
16. Fortier, "La Politique de l'espace parisien," 59.
17. J.-C. Perrot, *Genèse d'une ville moderne*, 12.
18. Favre, *La Mort dans la littérature*, 249.
19. *Etudes de la nature* (1784), 220–222, quoted in ibid., p. 250.
20. Jean-Noël Biraben, *Les Hommes et la peste en France et dans les pays européens et méditerranéens*, 3 vols. (Paris, 1975), 2:179.
21. Pierre Deyon, *Amiens, capitale provinciale* (Paris, 1967), 22.
22. Ibid., p. 27.
23. Detailed information can be found in Alphonse Chevallier, "Notice historique sur le nettoiement de la ville de Paris," *Annales d'Hygiène Publique et de Médecine Légale*, 42 (1849).
24. Chauvet, *Essai sur la propreté*, 28.
25. Tournon, *Moyen de rendre propres les rues*, 16.
26. Bertholon, *De la salubrité*, 90; Chauvet, *Essai sur la propreté*, 34.
27. Lavoisier, *Oeuvres*, 3:496.
28. Géraud, *Essai sur la suppression des fosses*, 58–59.
29. Pierre Saddy, "Le Cycle des immondices," *XVIIIe Siècle*, 9 (1977), 203–214; Arlette Farge, "L'Espace parisien au XVIIIe siècle d'après les ordonnances de police," *Ethnologie Française*, n.s., 12 (April–June 1982), 119–125.
30. See Saddy, "Le Cycle des immondices," 206.

31. For an exhaustive account see François-Joseph Liger, *Fosses d'aisances, latrines, urinoirs et vidanges* (1875), 342–384.
32. Ronesse, *Vues sur la propreté,* 31.
33. There is an excellent discussion of this subject in Favre, *La Mort dans la littérature,* 378 ff.
34. Changing the air, wrote Jean-Claude Perrot, "is not to assist the cure, it is positively to cure"; *Genèse d'une ville moderne,* 2:890.
35. Hales, *Description of Ventilators,* 69–76.
36. Géraud, *Essai sur la suppression des fosses,* 128.
37. François Béguin, "Evolution de quelques stratégies médico-spatiales," in Fortier, *La Politique de l'espace parisien,* 208.
38. On this subject see ibid., p. 228.
39. Samuel Sutton, *An Historical Account of a New Method for Extracting the foul air out of ships* (London, 1749), 13–16.
40. Hales, *Description of Ventilators,* 17–23.
41. De Genneté, *Purification,* 21.
42. Sutton, *New Method,* 4.
43. Laborie, Cadet the younger, and Parmentier, *Observations sur les fosses,* 26, 27, and 29.
44. Baumes, *Mémoire,* 186.
45. Ingenhousz, *Expériences sur les végétaux,* 162–163.
46. Howard, *Principal Lazarettos,* 100.
47. Banau and Turben, *Mémoire,* 53–57.
48. Baumes, *Mémoire,* 162.
49. Monfalcon, *Histoire des marais,* 384; a remedy that was not without its risks, since Monfalcon also stated (p. 126) that swampy emanations incited girls and women to licentious ways.
50. See Tournon, *Moyen de rendre propres les rues,* 24.
51. Mme. Marc-Amande Gacon-Dufour wrote in 1825 that "every traveler [confined in a carriage] is in his own interest obliged to have a flask of vinegar"; *Manuel du parfumeur,* 111.
52. Fortier, "La Politique de l'espace parisien," 60.
53. Navier, *Sur les dangers des exhumations,* 63.
54. Biraben, *Les Hommes et la peste,* 2:177.
55. Baumes, *Mémoire,* 163.
56. Banau and Turben, *Mémoire,* 68.
57. Guyton de Morveau, *Traité,* 7.
58. Banau and Turben, *Mémoire,* 55 ff. and 78.
59. Fortier, "La Maîtrise de l'eau."
60. Howard, *State of Prisons,* 264–266.
61. Biraben, *Les Hommes et la peste,* 2:170.
62. François Béguin, in Foucault et al., *Les Machines à guérir,* 40.
63. De Genneté, *Purification,* 24.
64. See Richard Etlin, "L'Air dans l'urbanisme des Lumières," *XVIIIᵉ Siècle,* 9 (1977), 123–134.
65. Just as many positive obsessions can be found throughout Howard's works.
66. Howard, *State of Prisons,* 22.

67. Baumes, *Mémoire,* 184.
68. J.-C. Perrot, *Genèse d'une ville moderne,* 2:686.
69. Ledoux cited in Ozouf, "L'Image de la ville."
70. Tenon, *Mémoires sur les hôpitaux,* 166.
71. See Etlin, "L'Air dans l'urbanisme," 132.
72. Jacquin, *De la santé,* 85 ff.
73. Géraud, *Essai sur la suppression des fosses,* 128.
74. Baumes, *Mémoire,* 184.
75. Etlin, "L'Air dans l'urbanisme," 132.
76. Ozouf, "L'Image de la ville," 1279. Ozouf provides an excellent analysis of Ledoux's exceptional talent.
77. Maurice Garden, *Lyon et les lyonnais au XVIIIᵉ siècle* (Lyons, 1970), 12.
78. Fortier, "La Politique dans l'espace parisien," 41 ff.
79. Ibid., p. 92.
80. Louis-René Villermé, *Des prisons telles qu'elles sont et telles qu'elles devraient être . . . par rapport à l'hygiène, à la morale et à l'économie politique* (1820), 39 ff.
81. Vigarello, *Le Corps redressé,* 123.
82. Jean-Louis Flandrin, *Familles, parenté, maison, sexualité dans l'ancienne société* (Paris, 1976), 97–101.
83. Philippe Perrot, *Les Dessus et les dessous de la bourgeoisie* (Paris, 1981), 288. Roche, *Le Peuple de Paris,* 133, points out that at the end of the eighteenth century everyone in the lodginghouse population had his own bed.
84. Tenon, *Mémoires sur les hôpitaux,* 165 ff.
85. See Michel Foucault, *Naissance de la clinique* (Paris, 1963), 38 ff., and Favre, *La Mort dans la littérature,* 246 ff.
86. Ariès, *L'Homme devant la mort,* 484 ff.
87. On the norms suggested by Maret (which he approved of) see Vicq d'Azyr, *Essai sur les lieux,* cxxix.
88. Thouret, *Rapport sur les exhumations du cimetière et de l'église des Saints-Innocents* (1789). The transfer took place from December 1785 to October 1787.
89. On this quest see Hallé, *Recherches sur la nature du méphitisme,* 10.
90. Biraben, *Les Hommes et la peste,* 2:176.
91. Navier, *Sur les dangers des exhumations,* 54.
92. Biraben, *Les Hommes et la peste,* 1:235.
93. Navier, *Sur les dangers des exhumations,* 52.
94. Lavoisier, *Oeuvres,* 3:477. This prescription was dictated by his theory of combustion. Note that aromatic fumigations combined the beneficial effects of fire with that of "perfumes."
95. Duhamel-Dumonceau, *Moyens de conserver la santé,* 119.
96. Thouret, *Rapport sur la voirie,* 7–8.
97. Apart, of course, from the virtues attributed to holy water, which was sprinkled on the ships of the Russian fleet struck by epidemic in 1795; Guyton de Morveau, *Traité,* 45.
98. Banau and Turben, *Mémoire,* 64.
99. Howard, *Principal Lazarettos,* 11.
100. "The cieling [*sic*] and walls of every ward and room should be well scraped; and then washed with the best stone-lime taken hot from the kiln, and slaked

in boiling water and size, and used during the strong effervescence; at least twice a year . . . Each ward and room should be swept, and *washed every day*, by the respective inhabitants; and sometimes with hot vinegar"; Howard, *State of Prisons*, 30. Howard then quotes from Lind's recommendations for fumigating ships, advising use of the same methods for prisons. After an infection, the premises should be fumigated with charcoal and brimstone every day for two weeks. After the final fumigation, "every thing ragged and dirty should be destroyed, as also the clothes and bedding of those who brought the infection . . . the bedding of such as have died of the fever, and unless the infection has been very mild, the bedding of such as have had the fever though recovered. The remaining clothes and bedding should be purified by being exposed twice a week to the steams of the brimstone and charcoal"; *State of Prisons*, 32.

101. Laborie, Cadet the younger, and Parmentier, *Observations sur les fosses*, 39.
102. Quoted in Thouret, *Rapport sur la voirie*, 14.
103. Navier, *Sur les dangers des exhumations*, 46.
104. Guyton de Morveau, *Traité*, 272.
105. Ibid., pp. 10–13.
106. Ibid., p. 13.
107. Vicq d'Azyr, *Instruction*, 7–8.
108. Guyton de Morveau, *Traité*, 93.
109. Ibid., p. 94.
110. Dr. James Carmichael-Smith, *Observations sur la fievre des prisons, sur les moyens de la prevenir . . . à l'aide des fumigations de gaz nitrique, et sur l'utilité de ces fumigations pour la destruction des odeurs et des miasmes contagieux* (1801), 88. Cruickshank, like Guyton de Morveau, used fumigations of oxygenated hydrochloric acid.
111. See Marcel Spivak, "L'Hygiène des troupes à la fin de l'Ancien Régime," *XVIIIᵉ Siècle*, 9 (1977), 115–122.
112. For this observation I am grateful to Jean Chagniot, an expert on the history of the French Guards at the end of the ancien régime.
113. Lind, *Essay on the Health of Seamen;* he was thinking primarily of the patients.
114. His strategy was summed up by Duhamel-Dumonceau, *Moyens de conserver la santé*, 73 ff.
115. Letter from John Haygarth, May 30, 1789, in Howard, *Principal Lazarettos*, app., p. 23.
116. Ibid. Identical concern can be noted on board the *Adventure*, under Captain Furneaux, Cook's companion; Captain Cook, *Three Voyages round the World*, ed. Lieutenant Charles R. Low (London, 1876), 313.
117. In Foucault et al., *Les Machines à guérir*.
118. Boissieu, *Dissertations sur les antiseptiques*, 66.
119. Hallé, "Air des hôpitaux," 575.
120. See the report by Delassone and Daubenton, June 20, 1787.
121. Etlin, "L'Air dans l'urbanisme," 132.
122. Hallé, "Air des hôpitaux," 575.
123. Howard, *Principal Lazarettos*, 37.
124. Saddy, "Le Cycle des immondices," 209.

125. Howard, *Principal Lazarettos,* 181, 182, 209.
126. Tenon, *Mémoires sur les hôpitaux,* 434.
127. See Chapter 7, pp. 124–125.
128. Lavoisier, *Oeuvres,* 3:469.
129. Howard, *Principal Lazarettos,* 218.
130. Title of the book by Geneviève Heller: *Propre en ordre. Habitation et vie domestique 1850–1930* (Paris, 1980).
131. Howard, *Principal Lazarettos,* 202–205.
132. Lavoisier, *Oeuvres,* 3:474 ff.

7. Odors and the Physiology of the Social Order

1. On systems of visual perception see Barthes, *Sade, Fourier, Loyola.*
2. Fourcroy and Berthollet referred to in Pierre Jean Robiquet, "Considérations sur l'arôme," *Annales de Chimie et de Physique,* 15 (1820), 28.
3. Locke had already suggested this theory; he adopted the Cartesian way of explaining the way in which the senses perceive the qualities of objects; *Essay Concerning Human Understanding,* 304–305.
4. On the *spiritus rector* theory, see Introduction, n. 14.
5. Lavoisier, "Mémoire sur les altérations qui arrivent à l'air dans plusieurs circonstances où se trouvent les hommes réunis en société," *Histoire et Mémoires de la Société Royale de Médecine,* 8 (1787), quoted in Félix Leblanc, *Recherches sur la composition de l'air confiné* (1842), 4.
6. Forget, *Médecine navale,* 191.
7. Piorry, *Des habitations,* 85 and 91.
8. Leblanc, *Recherches,* 7.
9. C. Grassi, *De la ventilation des navires* (1857), 5.
10. See Louis Chevalier, *Classes laborieuses et classes dangereuses à Paris pendant la première moitié du XIXᵉ siècle* (Paris, 1958), 168–182.
11. J.-B. Huzard the younger, *De l'enlèvement des boues et des immondices de Paris* (1826), urged the removal of the refuse dumps that were contaminating rue Château-Landon, rue de la Voirie, and the Montreuil, Fourneaux, and Enfer gates.
12. V. Moléon, *Rapports généraux sur les travaux du Conseil de Salubrité,* 2 vols. (Paris, 1828), 1:265 (report for 1823). See also R.-H. Guerrand, "Petite Histoire du quotidien. L'avènement de la chasse d'eau," *L'Histoire,* 43 (1982), 97. Witnesses' obsession with the danger from excrement was an extension of the eighteenth-century complaint, except that analysis of smells was becoming still more prominent. Many works attest the strength of the repulsion. Dr. Claude Lachaise, *Topographie médicale de Paris* (1822), 139 describes the odors from the Montfaucon rubbish dump; Dr. François-Marc Moreau, *Histoire statistique du choléra-morbus dans le quartier du faubourg Saint-Denis* (1833), 40, describes the Saint-Laurent fairground before its purification in 1832: "In many places the ground is so covered with fecal matter that it is no longer visible." Dr. Félix Hatin, *Essai médico-philosophique sur les moyens d'améliorer l'état sanitaire de la classe indigente . . .* (1832), probably best expresses the revulsion to the omnipresence of excrement, as in his description of the environs of Notre-

Dame (p. 3): "We civilized and refined people live amid an uncleanliness that is a constant reminder of the infirmities to which nature has condemned us from the cradle. Nothing is more shocking, in my opinion, than our great buildings edged with the residue of digestion."

Mrs. Trollope, *Paris and Parisians in 1835*, 1:113, claimed that "in a city as this, you are shocked and disgusted at every step you take, or at every gyration that the wheels of your chariot can make, by sights and smells that may not be described." Victor Considérant (see R.-H. Guerrand and Elsie Canfora-Argandona, *La Répartition de la population. Les conditions de logement des classes ouvrières à Paris au XIXᵉ siècle* [Paris, 1976], 19–20) and Balzac *(La Fille aux yeux d'or)* also provide evidence. Of all the authors I consulted, only Antoine Caillot, *Mémoires pour servir à l'histoire des moeurs et usages des français*, 2 vols. (1827), 1:303, congratulated himself on the decline in the fetidity of public space since the Consulate, and he referred to a very specific area: the gardens of the Palais-Royal. These were cleansed of their contaminating excrement at the end of the eighteenth century.

On the excremental hell of Lille, see Pierrard, *La Vie ouvrière*, especially his description of cesspool clearers who used barrels stopped up with a few blades of corn. The work was performed by the small "tosser" or "bernatier" merchants, who went through the streets with their carts crying "Four sous a barrel!" (p. 54), then delivered their goods to Flemish buyers. In 1850 Lille still did not possess public urinals; "a few buckets placed along the walls took their place; they were emptied near a cistern situated near the town hall" (p. 53).

On belated complaints about excrements at Nevers and La Charité-sur-Loire, see Guy Thuillier, *Pour une histoire du quotidien au XIXᵉ siècle en Nivernais* (Paris, 1977), 34.

13. See Chevalier, *Classes laborieuses*, 461–463.
14. See Alain Faure, "Classe malpropre, classe dangereuse?" *L'Haleine des Faubourgs, Recherches*, 29 (1977), 79–102.
15. Moléon, *Rapports généraux*, 2:46. This obsession with suburban contamination of the town center recurred at Lille in connection with the Sainte-Agnès rubbish dump (Pierrard, *La Vie ouvrière*, 53), where the peasants came and helped themselves.
16. This was an application of the principle of isolation, of separation, which inspired public health policy and particularly the medical code of March 8, 1822. See Blandine Barret-Kriegel, "Les Demeures de la misère," in Comité, *Politiques de l'habitat*, 93.
17. Moléon, *Rapports généraux*, 2:75.
18. A.-A. Mille, "Rapport sur le mode d'assainissement des villes en Angleterre et en Ecosse," *Annales d'Hygiène Publique et de Médecine Légale*, 2d ser., 4 (July–October 1855), 210, 209. He also asserted (p. 210) that the value of manure must be "calculated according to the absence of odor."
19. See Parent-Duchâtelet, *Rapport sur les améliorations*, 371.
20. Alexandre Parent-Duchâtelet, *Rapport sur les nouveaux procédés de MM. Salmon et Payen et Cie pour la dessication des chevaux morts . . .* (1833), reprinted in *Hygiène publique*, 2:293.
21. Dr. Emile-Louis Bertherand, *Mémoire sur la vidange des latrines et des urinoirs publics* (1858), 7.

22. H. Sponi, *De la vidange au passé, au présent et au futur* (1856), 29. The calculation from the *Journal de Chimie Médicale* is cited in Guerrand, "Petite Histoire du quotidien," 97.

23. Via his famous theory of the "circulus."

24. These concerns took root in the eighteenth century, when the manufacture of poudrette was established at Montfaucon.

25. Chevallier, "Notice historique," 318.

26. Pierre Pierrard has noted that during the Second Empire the old and infirm still accounted for half the manpower employed by rubbish contractors.

27. Chevallier, "Notice historique," 307, 319, and 313; the italics are mine.

28. See Bertherand, *Mémoire*, passim, already cited by Laporte, *Histoire de la merde*.

29. Sponi, *De la vidange*, 26.

30. Laporte, *Histoire de la merde*, 99 ff.

31. On these fluctuations see Liger, *Fosses d'aisances*, 87 ff. The author provides some information on the evolution of prices.

32. Moléon, *Rapports généraux*, 2:234 (report for 1835). Separating apparatus inspired a copious body of literature.

33. Pierrard, *La Vie ouvrière*, 49.

34. Gabriel Désert, *Histoire de Caen* (Paris, 1981), 199 and 228.

35. Thuillier, *Pour une histoire du quotidien*, 34.

36. "Rapport d'Emile Trélat sur l'évacuation des vidanges hors des habitations," in *De l'évacuation des vidanges dans la ville de Paris* (1880–82), 29 (read January 25, 1882).

37. Gérard Jacquemet, "Urbanisme parisien: La bataille du tout-à-l'égoût à la fin du XIXᵉ siècle," *Revue d'Histoire Moderne et Contemporaine*, 26 (October–December 1979), 505–548.

38. Chevreul, "Mémoire," 42.

39. Hippolyte Marié Davy, in *De l'évacuation des vidanges*, 67 ff. This content varied between 9 kilograms per cubic meter for material that was not mixed with water, collected from the premises of the masses, and 270 grams per cubic meter in the cesspool of the Grand Hôtel.

40. Parent-Duchâtelet, fervent propagandist for the product, advised organizing demonstrations on the pavements of the city in order to prove its efficiency to the public; *Rapport sur les améliorations*, 397.

41. Georges Knaebel, *Les Problèmes d'assainissement d'une ville du Tiers Monde: Pointe Noire* (Thesis 3d cycle, October 1978), 249.

42. Alexandre Parent-Duchâtelet, *Les Chantiers d'équarrissage de la ville de Paris envisagés sous le rapport de l'hygiène publique* (1832), 29, 100.

43. Lachaise, *Topographie médicale de Paris*, 139.

44. Moléon, *Rapports généraux*, 1:89 (report for 1815).

45. See Parent-Duchâtelet, *Les Chantiers*, 28.

46. This was the view of J.-B. Monfalcon and A. Polinière, who calculated the profits of this industry, *Traité de la salubrité dans les grandes villes* (1846), 220 ff.

47. See Moléon, *Rapports généraux*, 2:16 (report for 1827).

48. Ibid., 1:325 (report for 1825).

49. Ibid., p. 286 (report for 1824).

50. Parent-Duchâtelet, *Rapport sur les nouveaux procédés*, 295, and *Projet . . . d'un*

rapport . . . sur la construction d'un clos central d'équarrissage pour la ville de Paris, reprinted in *Hygiène publique,* 2:310.

51. Monfalcon and Polinière, *Traité de la salubrité,* 224.
52. Alexandre Parent-Duchâtelet, *De l'influence et de l'assainissement des salles de dissection,* reprinted in *Hygiène publique,* 2:22–24.
53. Jean Chrétien, *Les Odeurs de Paris* (1881), 33.
54. Widely practiced by Western doctors. See Jacques Léonard, *Les Médecins de l'Ouest au XIXᵉ siècle,* vol. 3 (Paris, 1979), 1141.
55. Reutter de Rosemont, *Histoire de la pharmacie,* 286.
56. Antoine-Germain Labarraque, *Observations sur l'emploi des chlorures* (1825), 5; text of Delavau's report reproduced ibid., p. 3.
57. Comment by Labarraque reported in a letter by Maxime du Camp and published in *La Chronique Médicale,* 1915, p. 280.
58. Parent-Duchâtelet, *Rapport sur le curage,* 362.
59. Nicolas-Michel Troche, *Notice historique sur les inhumations provisoires faites sur la place du marché des Innocents en 1830* (1837); and Alexandre Parent-Duchâtelet, *Note sur les inhumations et les exhumations qui ont lieu à Paris, à la suite des événements de juillet 1830,* reprinted in *Hygiène publique,* 2:81.
60. *Mémoires de M. Gisquet,* vol. 1 (1840), 425–427. See also Barret-Kriegel, "Les Demeures de la misère," 108.
61. See Chapter 2, pp. 30–31, and Moléon, *Rapports généraux,* 1:264 (report for 1823).
62. Alexandre Parent-Duchâtelet and Jean-Pierre d'Arcet, *De l'influence et de l'assainissement des salles de dissection* (1831).
63. Quoted in Labarraque, *Observations,* 5.
64. Moléon, *Rapports généraux,* 2:428 (report for 1838).
65. Honoré de Balzac, *Un Début dans la vie* (Paris, 1976), 777.
66. Sponi, *De la vidange,* 8.
67. The names of Boussingault, d'Arcet, Dupuytren, Fourcroy, Hallé, Labarraque, and, in the bibliography, of Parent-Duchâtelet, Parmentier, Payen, Thouret, and Trébuchet are particularly noteworthy. On this subject see ibid., p. 10.
68. Ibid., p. 27.
69. Thomas Tredgold, *Principes de l'art de chauffer et d'aérer les edifices publics, les maisons d'habitation, les manufactures, les hôpitaux, les serres . . .* (1825); first published as *The principles of warming and ventilating public dwellings, dwelling houses . . .* (London, 1824). Thus Maurice Daumas's statement (*Histoire générale des techniques,* vol. 3 [Paris, 1969], 522–523) that ventilation did not vary during this period requires qualification.
70. Philippe Grouvelle, Introduction to Jean-Pierre d'Arcet, *Collection de mémoires relatifs à l'assainissement des ateliers, des édifices publics et des maisons particulières,* vol. 1 (1843), vii.
71. Tredgold, *Principes des chauffer,* 271.
72. D'Arcet, "Rapport sur des . . . fourneaux de cuisine salubres et économiques" (1821), in *Collection de mémoires,* 113.
73. It is not my intention to analyze how these same plans recur in social perceptions; but we know how they guided the desire to control prostitution in total isolation.

74. Philippe Grouvelle, Introduction to d'Arcet, *Collection de mémoires*, vi.
75. Villermé, *Des prisons*, 18.
76. *Chauffage et ventilation de la Nouvelle Force par Philippe Grouvelle* (1845), 25.
77. See Leblanc's *Recherches* for a synthesis of experiments carried out in a bedroom, an infant school, a primary school classroom, a lecture room in the Sorbonne, the Chamber of Deputies, an entertainment hall (salle Favart), the military stables, and a greenhouse in the Jardin Royal. In each of these places Leblanc noted "the capacity of the enclosed area, the number of individuals, the period of time it was closed, the temperature, the method of heating, the absence or existence . . . of ventilation," measured by Combes's anemometer (p. 11). For the analysis of confined air, see also Eugène Péclet, *Instruction sur l'assainissement des écoles primaires et des salles d'asile* (1846).
78. See Dr. C. Grassi, *Rapport . . . sur la construction et l'assainissement des latrines et fosses d'aisances* (1858), 32.
79. Edouard Ducpétiaux, "Extrait du rapport sur les deux systèmes de ventilation établis à titre d'essai dans la prison cellulaire des femmes, à Bruxelles," *Annales d'Hygiène Publique et de Médecine Légale*, 50 (1853), 459 ff.
80. Ibid., p. 461.
81. Ibid.
82. Grassi, *De la ventilation*, 23.
83. Geneviève Carrière and Bruno Carrière, "Santé et hygiène au bagne de Brest au XIXe siècle," *Annales de Bretagne et des Pays de l'Ouest*, 3 (1981), 349. In 1822 the engineer Trotté de la Roche wrote, concerning the penal colony at Brest, "at night men do not bother to go to the latrines for their minor needs. Instead of entering the aqueduct, the urine stays on the floor and impregnates the wood."
84. See Chapter 10, pp. 173–175.
85. See Dominique Laporte, "Contribution pour une histoire de la merde: La merde des asiles, 1830–1880," *Ornicar*, 4 (July 1977), 31–48.
86. Quoted in Grassi, *Rapport sur la construction*, 37.
87. Edmond Duponchel, "Nouveau système de latrines pour les grands établissements publics et notamment pour les casernes, les hôpitaux militaires et les hospices civils," *Annales d'Hygiène Publique et de Médecine Légale*, 2d ser., 10 (July– October 1858), 356–362.
88. Quoted by Grouvelle, Introduction to d'Arcet, *Collection de mémoires*, xxiii.
89. François Caron, *Histoire économique de la France, XIXe–XXe siècle* (Paris, 1981), 65.

8. Policy and Pollution

1. See Conseil de Salubrité, Receuil des Plaintes, Archives de la Préfecture de Police, Paris.
2. Piorry, *Des habitations*, 38.
3. Arlette Farge, "Les Artisans malades de leur travail," *Annales. Economies, Sociétés, Civilisations*, 32 (September–October 1977), 993–1006.
4. Report by the minister of the interior concerning the motives of the decree

of October 15, 1810; quoted in Dr. Maxime Vernois, *Traité pratique d'hygiène industrielle et administrative,* vol. 1 (1860), 14.

5. Quoted in ibid., p. 28.

6. Moléon, *Rapports généraux,* 2:iv.

7. Dangerous or unhealthy establishments, buildings for carrying on noisy or noxious trades, and the rest.

8. Report by the Institut's Section de Chimie, presented in 1809 to the Classe des Sciences Physiques et Mathématiques; quoted in Vernois, *Traité pratique,* 18.

9. Monfalcon and Polinière, *Traité de la salubrité,* 172.

10. See B.-P. Lécuyer, "Démographie, statistique et hygiène publique sous la Monarchie censitaire," *Annales de Démographie Historique,* 1977, p. 242.

11. On barracks see Moléon, *Rapports généraux,* 2:123 ff.; on prisons see ibid., pp. 141–150 (report for 1829).

12. See ibid., p. 185 (report for 1821), and Alexandre Parent-Duchâtelet, *Recherches et considérations sur la rivière de Bièvre ou des Gobelins, et sur les moyens d'améliorer son cours* (1822).

13. *Classes laborieuses,* 173 ff.

14. This intolerance came later in the provinces: at Nevers, complaints about black dust became numerous from 1854 on; Thuillier, *Pour une histoire du quotidien,* 38–39.

15. Charles de Launay (Mme. Emile de Girardin) complained bitterly in 1837 about this ubiquitous odor: "Every minute you are stifled by an objectionable odor . . . at the corner of every boulevard you see enormous boilers on large fires poked by little men with strange faces"; Letter XIX, *Lettres parisiennes* (1843), 181. Dominique, when he first comes to Paris, is struck by the strong odor of gas; Eugène Fromentin, *Dominique* (Paris, 1972), 132 (1st ed. 1862).

16. Attempts to reduce smoke and dust were made at Fourchambault even before 1850; Thuillier, *Pour une histoire du quotidien,* 35.

17. Monfalcon and Polinière, *Traité de la salubrité,* 327–351.

18. The Commission des Logements Insalubres in Paris confirmed both this primacy of smell and the advent of new considerations. When its members took office in November 1850, they posed one preliminary question: "What should be understood by insalubriousness? . . . in that respect it agrees with the Conseil de Salubrité that there is insalubriousness everywhere there is an unpleasant smell capable of vitiating the air in habitations, everywhere humidity and uncleanliness prevail, where there is lack of air and light"; Département de la Seine, Ville de Paris, Commission des Logements Insalubres, *Rapport général des travaux de la Commission . . . pendant l'année 1851* (Paris, 1852), 4.

19. This concern emerged in 1847; Moléon, *Rapports généraux,* 2:1075 ff.

20. Léonard, *Les Médecins,* 1151.

21. Knaebel, *Les Problèmes d'assainissement,* 242–243, and Gabriel Dupuy and Georges Knaebel, *Choix techniques et assainissement urbain en France de 1800 à 1977* (Paris, 1979). In Haussmann's eyes, according to Knaebel (p. 242), there was the city, to be embellished in those places where the bourgeois gave himself to the enjoyment of perception, where nothing must offend the senses— which implied the expulsion of the dirty, the poor, the unclean, the mal-

odorous—and the "noncity." In Knaebel's view, it is in this context that Haussmann envisaged an underground sewer system. The interpretation is stimulating; and the statement that the method of evacuation of excrement only translates the shape of the ratio of social power (p. 46) deserves even more precise analysis. Nevertheless, work by historians, particularly by Jeanne Gaillard and Jean Le Yaouanq, has shown how the traditional town resisted, often successfully, the civic enterprises of the Second Empire. The expulsion of the poor and the marginal from the center was not nearly as obvious as is too often claimed; and, as we have seen, mains drainage triumphed only at the end of the century. In short, apart from the VIIIe arrondissement, the noncity remained established in the heart of the city.

22. On this subject Daniel Roche, in a paper presented at the Franco-Quebec colloquium (Ecole des Hautes Etudes en Sciences Sociales, May 1981), commented that the complaints about nuisances were not identical in different districts. Parent-Duchâtelet referred to this sociology of thresholds of tolerance in a quite different context when he wrote that a certain brothel, considered scandalous in the rue Feydeau, would pass totally unnoticed in a "lowly" district; *La Prostitution à Paris au XIXe siècle* (Paris, 1981), 150 (extracts first published in 1836 under the title *De la prostitution dans la ville de Paris . . .*).

Introduction to Part Three

1. Cabanis, *Rapports du physique,* 526, 527, 528. This theory followed Maine de Biran's distinction between purely passive sensation and perception that presupposed some activity by the organs. In contrast, Destutt de Tracy regarded perception as a specific sensation, spread out before the mind. See Jean-Pierre Richard, *Littérature et sensation* (Paris, 1963), 28 and 112.

2. Wilhelm Fliess, *Les Relations entre le nez et les organes génitaux féminins présentés selon leur signification biologique* (Paris, 1977; 1st ed. 1897).

3. Cabanis, *Rapports du physique,* p. 102.

4. "For more than half a century," Havelock Ellis observed, "no important progress was made in this field . . . the subject of smell was mainly left to those interested in 'curious' subjects"; *Sexual Selection in Man,* 51.

5. "There is no connection between the nature of the sensory principle and the nature of understanding," Tourtelle proclaimed peremptorily in 1815; *Eléments d'hygiène,* 479.

6. Cabanis, *Rapports du physique,* 293, 543 ff.

7. Quoted in Cloquet, *Osphrésiologie,* 45.

8. Virey, "Des odeurs," 256. The author's comments on sensory acuteness, apprenticeship, and behavior form part of the anthropologists' research program, defined by Joseph-Marie de Gerando; see Jean Copans and Jean Jamin, *Aux origines de l'anthropologie française* (Paris, 1981), 149.

9. Kirwan, *De l'odorat,* 32–34.

10. Virey, "Des odeurs," 256. He was inspired by Cook's observations. The theme was revived by Cloquet, *Osphrésiologie,* 137. Alexander, indefatigable collector

of refuse in Michel Tournier's *Météores*, was an excellent analyst of foul odors.

11. Michel Lévy, *Traité d'hygiène*, 2 vols. (1856), 1:91.

9. The Stench of the Poor

1. P.-A. Piorry, "Extrait du rapport sur les épidémies qui ont régné en France de 1830 à 1836, lu le 9 août 1836," *Mémoires de l'Académie Royale de Médecine*, 6 (1837), 17.
2. Passot, *Des logements insalubres*, 26.
3. See Maurice Agulhon, *Le Cercle dans la France bourgeoise, 1810–1848. Etude d'une mutation de sociabilité* (Paris, 1977), 79.
4. Surely Madame de Girardin was insinuating this when she somewhat pessimistically wrote on October 21, 1837: "Those who do not wash their hands will always hate those who wash their hands, and those who wash their hands will always despise those who do not wash their hands. You will never be able to bring them together, they will never be able to live together . . . because there is one thing which can not be overcome and that is disgust; because there is another thing that can not be tolerated, and that is humiliation"; de Launay, *Lettres parisiennes*, 190.
5. Charles-Léonard Pfeiffer, *Taste and Smell in Balzac's Novels* (Tucson, 1949).
6. Victor Hugo, *Les Misérables*, 2 vols. (Paris, 1963), 2:513.
7. Passot, *Des logements insalubres*, 26. However, if one goes by the volume of publications, the censitaire monarchy remains the golden age of neo-Hippocratic "medical topography."
8. Quoted in Dr. Henri Bayard, *Mémoire sur la topographie médicale du IVᵉ arrondissement de Paris* . . . (1842), 103 ff.
9. Except in terms of food; see Jean-Paul Aron, *Le Mangeur au XIXᵉ siècle* (Paris, 1976).
10. *Les Misérables*, 2:512.
11. Howard, *Principal Lazarettos*, 25.
12. Which Ramel advised, *De l'influence de marais*, 271–272.
13. Hallé, "Air des hôpitaux," 571.
14. In some passages Louis-Sébastien Mercier anticipated the tone of later descriptions. An example is his appalled recoil before the animality that prevailed in the faubourg St.-Marcel (see Roche, *Le Peuple de Paris*, 100). Nevertheless Roche recognized that medicine was at the time encroaching on private life.
15. Chauvet, *Essai sur la propreté*, 10.
16. Ibid., p. 8. This theme was expanded at length in Spain in the Golden Century; see Lapouge, "Utopie et hygiène," 117.
17. Ramazzini, *De morbis artificum*, 383.
18. Malouin, *Chimie médicinale*, 55.
19. C.-F. Hufeland, *La Macrobiotique ou l'art de prolonger la vie de l'homme* (1838), 472 (1st ed. in German, 1797); domestic servants, chamber pots, and linen drying round the stove all came under this proscription from the nursery. These social perceptions, however, did not prevent improvement in the status of domestic servants; see Roche, *Le Peuple de Paris*, 76 ff.
20. In my introduction to the 1981 edition of Parent-Duchâtelet, *La Prostitution à Paris au XIXᵉ siècle*.

21. See Jean-Jacques Darmon, "Sous la Restauration, des juges sondent la plaie si vive des prisons," in *L'Impossible Prison*, ed. M. Perrot (Paris, 1979), 123–146; and Hélène Chew, "Loin du débat pénitentiaire: La prison de Chartres durant la première moitié du XIXᵉ siècle," *Bulletin de l'Institut d'Histoire de la Presse et de l'Opinion* (Tours) 6 (1981), 43–67.

22. Quoted in Villermé, *Des prisons*, 25 and 26.

23. Moléon, *Rapports généraux*, 1:225. There are also innumerable other references to the stench and foul impregnation of the ragpicker. Examples include Moreau, *Histoire statistique*, 41; Lachaise, *Topographie médicale de Paris*, 190–192; Commission des Logements Insalubres, *Rapport général des travaux pendant l'année 1851*, 12; Passot, *Des logements insalubres*, 3. On the ragpickers of Lille see Pierrard, *La Vie ouvrière*, 54.

24. Report by the sanitary commission of the Jardin des Plantes, November 8, 1831, quoted in *Annales d'Hygiène Publique et de Médecine Légale*, 7 (1832), 200.

25. Barret-Kriegel, "Les Demeures de la misère," 130.

26. See Jean-Paul Aron and Roger Kempf, "Canum more," in *Le Pénis et la démoralisation de l'Occident* (Paris, 1978), 47 ff.

27. "The odor exhaled by these sorts of places is one of the circumstances which a very numerous category of pederasts seek, as it is indispensable to their pleasures"; Félix Carlier, *Etudes de pathologie sociale. Les deux prostitutions* (1887), 305 and 370. This part of the book has just been reissued under the title *La Prostitution antiphysique* (Paris, 1981).

28. Forget, *Médecine navale*, 127.

29. Dr. J.-M.-G. Itard, *Premier rapport . . . sur le sauvage de l'Aveyron* (1807), 88. Itard attributed this indifference to lack of education of the senses. His report has recently been reissued by Thierry Gineste in *Victor de l'Aveyron, dernier enfant suavage, premier enfant fou* (Paris, 1981). See also Harlan Lane, *The Wild Boy of Aveyron* (Cambridge, Mass., 1976).

30. Forget, *Médecine navale*, 126.

31. Ibid., p. 128.

32. Ibid., p. 135.

33. See Passot, *Des logements insalubres*, 7.

34. Gustave Flaubert, *Selected Letters*, trans. and ed. Francis Steegmuller (London, 1959).

35. Léonard, *Les Médecins*, 1140.

36. Quoted in Pierre Arches, "La Médicalisation des Deux-Sèvres au milieu du XIXᵉ siècle," *Bulletin de la Société Historique et Scientifique des Deux-Sèvres*, 3d quarter 1979, p. 261.

37. Jules Vallès, *L'Enfant* (Paris, 1972), 65.

38. Quoted in Pierrard, *La Vie ouvrière*, 87.

39. Thierry Leleu, "Scènes de la vie quotidienne: Les femmes de la vallée de la Lys: 1870–1920," *Histoire des Femmes du Nord, Revue du Nord*, 63 (July–September 1981), 661.

40. Marie-Hélène Zylberberg-Hocquard, "L'Ouvrière dans les romans populaires du XIXᵉ siècle," ibid., p. 629.

41. This is too vast a subject to deal with here. See Ned Rival, *Tabac, miroir du temps. Histoire des moeurs et des fumeurs* (Paris, 1981).

42. Théodore Burette, *La Physiologie du fumeur* (1840), 21.
43. See Agulhon, *Le Cercle*, 53. This is clearly the opinion of L. Rostan, *Cours élémentaire d'hygiène*, 2d ed., 2 vols. (1822), 1:546 ff.
44. Michelet, *Histoire de France*, 11:285–287; Adolphe Blanqui, *Des classes ouvrières en France pendant l'année 1848* (1849), 209.
45. Forget, *Médecine navale*, 294.
46. Burette, *Physiologie du fumeur*, 86.
47. Ibid., p. 79.
48. Ibid., p. 75.
49. Parent-Duchâtelet's attitude is significant from the point of view of the fear of infection. Here are two examples of the precautions advised when visiting the sick. Fodéré recommended: "One must keep one's clothes entirely buttoned up . . . one must never swallow one's saliva; one must spit and blow one's nose every time the need arises and, as in hospitals, wear an apron on which one frequently wipes one's hands . . . after having had the covers raised, one must wait a few moments before bending down and breathing [the invalid's] first emanations; moreover, one must always avoid his breath and keep at a reasonable distance from his mouth"; *Traité de médecine légale*, 6:111. A policy of keeping at a distance from foul bodies was formulated in this way. Tireless visitor to prisons, lazarettos, and hospitals, Howard confessed that he always avoided standing to leeward of the invalid. He constantly strove to hold his breath as much as possible; *State of Prisons*, 451; *Principal Lazarettos*, 232.
50. Monfalcon and Polinière, *Traité de la salubrité*, 90.
51. Dr. A. Joiré, "Des logements du pauvre et de l'ouvrier considérés sous le rapport de l'hygiène publique et privée dans les villes industrielles," *Annales d'Hygiène Publique et de Médecine Légale*, 45 (1851), 310.
52. Blanqui, *Des classes ouvrières*, 103, 98.
53. Foucault, *Naissance de la clinique*, 167.
54. Paul Gerbod, *La Condition universitaire en France au XIXe siècle* (Paris, 1965), 629.
55. Revealing on this subject was the foul odor of Brother Archangias in Zola's *La Faute de l'abbé Mouret*.
56. Norbert Truquin, *Mémoires, vie, aventure d'un prolétaire à travers la révolution* (Paris, 1977), 129 (1st ed. 1888). (The description applies to 1852.) On these fringe workers, see Jacques Rancière, *La Nuit des prolétaires* (Paris, 1981).
57. Passot, *Des logements insalubres*, 16.
58. Zylberberg-Hocquard, "L'Ouvrière dans les romans populaires," 627–628. See especially the descriptions of the Lille cellars and courtyards in Mathilde Bourdon's *Euphrasie, histoire d'une femme pauvre* (1868) and in M.-L. Gagneur's *Les Réprouvées* (1867).
59. Blanqui, *Des classes ouvrières*, 71.
60. Bayard, *Mémoire sur la topographie médicale*, 49.
61. Cf. Lachaise, *Topographie médicale de Paris*, 198. But the *Rapports du Conseil de Salubrité de la Seine* reveal an increase in public health concern regarding the presence of animals in Paris; dairy farms (1810–1820) and piggeries (1849–1858) first drew attention; after 1859 the complaints became more diffuse;

the desire was to stem the presence of animals generally. In 1880 there were complaints about the odor emanating from the hospital for sick dogs.

62. Piorry, "Extrait du rapport," 17.

63. Chevalier, *Classes laborieuses,* 182.

64. See Alain Corbin, "Les Paysans de Paris," *Ethnologie Française,* 2 (1980), 169–176.

65. Martin Nadaud, *Mémoires de Léonard, ancien garçon maçon,* ann. Maurice Agulhon (Paris, 1976), 103; O. d'Haussonville, "La Misère à Paris. La population nomade, les asiles de nuit et la vie populaire," *Revue des Deux-Mondes,* 47 (October 1881), 612; Pierre Mazerolle, *La Misère de Paris. Les mauvais gîtes* (1874), 28–31. Was it by chance that early nineteenth-century gastronomy tended to ignore cheese?

66. See the report on lodgings in *Statistique de l'industrie à Paris résultant d'une enquête faite par la Chambre de Commerce pour les années 1847–1848* (1851).

67. Victor Hugo, *Les Travailleurs de la mer* (Paris, 1980), 220.

68. Piorry, "Extrait du rapport," 17.

69. Quoted in Jean Borie, *Mythologies de l'hérédité au XIXᵉ siècle* (Paris, 1981), 113.

70. See, for example, Joire, "Des logements du pauvre," 318.

71. Ibid., p. 320.

72. Piorry, *Des habitations,* 74.

73. Jean Starobinski, "Sur la chlorose," *"Sangs,"* special issue of *Romantisme,* 1981, pp. 113–130.

74. Joiré, "Des logements du pauvre," 296.

75. Jules Michelet, *La Femme* (Paris, 1981), 90 (1st ed. 1859).

76. Cervantes, *Don Quixote,* part 1 (Paris, 1946), 219.

77. Ramazzini, *De morbis artificum,* 447–448.

78. See Chapter 5, p. 78.

79. And even beyond; see Rose-Marie Lagrave, *Le Village romanesque* (Le Paradou, 1980).

80. See Neil MacWilliams, paper presented at a colloquium at Loughborough University, September 1981. On the other hand, ethnographic projects at the beginning of the century gave up the study of material life and ecologicosocial observation; they neglected the material anthropology initiated by medical topographies. See Mona Ozouf, "L'Invention de l'ethnographie française: Le questionnaire de l'Académie celtique," *Annales. Economies, Sociétés, Civilisations,* 36 (March–April 1981), 213.

81. See Henry Roberts, *The Dwellings of the Working Classes* (London, 1850). Arthur Young also likened the peasants of Combourg to boors (*Travels in France,* 97). The metaphor that was to weigh heavily on rural historiography took root at that time; traces of it can be found in the recent very interesting book by Eugen Weber, *Peasants into Frenchmen* (Stanford, 1976).

82. Balzac, *Les Paysans* (Paris, 1978), 121.

83. See Alain Corbin, *Archaïsme et modernité en Limousin au XIXᵉ siècle,* vol. 1 (Paris, 1975), 74–94, for a discussion of the poor hygiene of the peasants in the Limoges region in the mid-nineteenth century. Also revealing is Guy Thuillier, *Aspects de l'économie nivernaise au XIXᵉ siècle* (Paris, 1966).

84. See Laporte, *Histoire de la merde*, 42.
85. Agrarian ideology is analyzed by Pierre Barral, *Les Agrariens français de Méline à Pisani* (Paris, 1968).
86. Piorry, "Extrait du rapport."
87. Thuillier, *Pour une histoire du quotidien*, 64, points out that belief in the uselessness of hygienic installations for country people persisted in the Nivernais until at least the early twentieth century.
88. The stinking promiscuity of military barracks remained the paradigm of repulsive smells for the young bourgeois. It was this that convinced the newly recruited Pierre Louÿs that he had to get himself discharged. (Paul-Ursin Dumont has kindly passed on to me Louÿs's unpublished correspondence.)
89. Quoted in Monin, *Les Odeurs du corps humain,* 72.
90. See Carl Vogt, *Leçons sur l'homme* (1865). He wrote (p. 161): "The exhalations from the skin also have their specific characteristics, which in certain races do not disappear in any circumstances, even the most scrupulous cleanliness. These characteristic odors of race should in no way be confused with the exhalations that originate from type of food, and can be noted in the same race . . . the specific odor of the Negro remains the same whatever attention he pays to cleanliness or whatever food he takes. It belongs to the species as musk does to the musk deer that produces it."
91. Blanqui, *Des classes ouvrières,* 151.
92. Luc Boltanski, *Prime Education et morale de classe* (Paris, 1969), 110.
93. Moléon, *Rapports généraux,* 1:199.
94. Gustave de Gérando, *Le Visiteur du pauvre,* 3d ed. (1826), 227.
95. Monfalcon and Polinière, *Traité de la salubrité,* 91, 89.
96. Emile Zola, *La Joie de vivre* (Paris, 1975), 1026.
97. See P. Perrot, *Les Dessus,* 227.
98. Cadet de Vaux, "De l'atmosphère," 435.
99. Passot, *Des logements insalubres,* 20.
100. Ibid., p. 21.
101. See Chapter 13, pp. 212–215.
102. See Piorry, *Des habitations,* 93.
103. Dr. Louis-René Villermé, "Sur les cités ouvrières," *Annales d'Hygiène Publique et de Médecine Légale,* 43 (1850), especially 246–258. See also, for example, Guerrand and Confora-Argandona, *La Répartition,* 33–41.
104. Gisquet, *Mémoire,* 1:423–424.
105. Mille, "Rapport sur la mode d'assainissement," 223.
106. Ibid., p. 213.
107. That is how he expressed himself in the Chamber of Deputies on July 13, 1848, in supporting the decree introduced by Emery on July 12. Note that these discussions took place two weeks after the crushing of the June uprising. On July 17 Anatole de Melun proposed the law that was the subject of the Riancey report, read on December 8, 1849.
108. Dr. Moreau had already compiled a house-to-house report at the time of the inquiry in Paris after the cholera morbus epidemic. In a more general way, Blandine Barret-Kriegel ("Les Demeures de la misère," 119 ff.) is right to regard this episode as a great turning point in the history of research techniques.

109. Monfalcon and Polinière, *Traité de la salubrité,* 92.
110. Passot, *Des logements insalubres,* 20.
111. For Paris, see Guerrand, "Petite Histoire du quotidien," 55 ff.; A. Thalamy, in Comité, *Politiques de l'habitat,* 59; and especially Danielle Rancière, "La Loi du 13 juillet 1850 sur les logements insalubres. Les philanthropes et le problème insoluble de l'habitat du pauvre," in ibid., pp. 187–207. For Lille, see Pierrard, *La Vie ouvrière,* 92 ff. For the poor application of the law in the Nivernais, see Thuillier, *Pour une histoire du quotidien,* 36 ff.

10. *Domestic Atmospheres*

1. On Claude-Nicolas Ledoux, see Ozouf, "L'Image de la ville," 1279–80.
2. Jacquin, *De la santé,* 294–295.
3. Mauzi, *L'Idée du bonheur,* 281.
4. Bayard, *Mémoire sur la topographie médicale,* 90.
5. Quoted in Chevalier, *Classes laborieuses,* 179.
6. Erving Goffman, *La Mise en scène de la vie quotidienne,* vol. 2 (Paris, 1973), 62.
7. Quoted in Passot, *Des logements insalubres,* 16.
8. Lévy, *Traité d'hygiène,* 1:544.
9. Ibid., p. 545.
10. Ibid.
11. P. 131.
12. Mille, "Rapport sur la mode d'assainissement," 199.
13. Jules Michelet, *Histoire de la Régence* (1863), 394.
14. Piorry, *Des habitations,* 126.
15. Ibid., p. 57.
16. Hufeland, *La Macrobiotique,* 470.
17. Louis Odier, *Principes d'hygiène extraits du code de santé et de longue vie de Sir John Sinclair* (1823), 574.
18. Londe, *Nouveaux Eléments,* 1:405 ff.
19. Odier, *Principes d'hygiène de Sinclair,* 577.
20. Anne Martin-Fugier, *La Place des bonnes. La domesticité féminine à Paris en 1900* (Paris, 1979), 113.
21. Piorry, *Des habitations,* 85.
22. Both Piorry (*Des habitations,* 104), and Londe (*Nouveaux Eléments,* 2:322) denounced the intolerable odor from stoves. Thuillier, *Pour une histoire du quotidien,* 41.
23. Writing about the Nivernais, Thuillier stressed the women's attachment to footwarmers and their refusal to replace them with hotwater bottles; ibid., p. 48. See also Cabanès, *Moeurs intimes,* 67 ff.
24. "A foul odor," according to Rostan, *Cours élémentaire d'hygiène,* 2:44.
25. An act that revealed a new sensitivity. According to J.-P. Chaline, *La Bourgeoisie rouennaise au XIXᵉ siècle* (Thesis Paris IV, 1979), 805, "Three things are forbidden in the chamber: perfumes, toilet waters, and shoes, all because of their odors."
26. Fodéré (*Traité de médecine légale,* 5:44) continued to demand individual cradles for the nurselings crowded into the foundlings' hospital at Marseilles and boasted of "the order established in every lycée in the French Empire, where

each pupil has a separate bedroom, but not a separate ceiling, so that not only does the air circulate freely from all sides, but also the pupil can be supervised at every moment of day and night" (5:48). The real problem was to achieve the necessary subtle balance: to isolate odors without impeding ventilation, to abolish promiscuity and therefore homosexual relationships while at the same time controlling masturbation.

27. Londe, *Nouveaux Eléments,* 1:404.

28. And proved it by drawing up a meticulous catalog of references to olfaction in *La Comédie humaine;* Pfeiffer, *Taste and Smell,* 100–114.

29. For the clerk, "the atmospheric circumstances consisted of the air of the corridors, the masculine exhalations contained in unventilated chambers, the smell of pens and pencils"; Balzac, *Physiologie de l'employé* (1841), 44. Emile Gaboriau, in *Les Gens de bureau* (1862), allotted a large place to odors in his narrative. See Guy Thuillier, *La Vie quotidienne dans les ministères au XIXᵉ siècle* (Paris, 1976), 15, 16, and 41. Regional odors obtruded here as in the dormitories: "There is the Alsatians' office, which smells of sauerkraut, and the Provencals' office, which smells of garlic" (p. 41). The large increase in female employees around 1900 transformed the smell of the atmosphere in offices. Cheap perfume and flowers relieved its rancid atmosphere, causing what Guy Thuillier called "the unpleasant odors of the 1880s" to disappear. Earlier complaints had associated the stench of offices with masculinity and celibacy; we have seen why.

30. The stench of the courtroom, where criminal and stinking poverty were on show to the elite, avid for strong sensations, continued to be a common theme; this emphasis was the remote legacy of the terror inspired by "jail fever." See Jean-Louis Debré, *La Justice au XIXᵉ siècle. Les magistrats* (Paris, 1981), 176.

31. Balzac, *Old Goriot,* trans. M. A. Crawford (Harmondsworth, 1981), 31. The stench of the college boardinghouse was even worse (see *Louis Lambert,* passim). The importance of the smells of this environment in the genesis of male sensitivity in the nineteenth century cannot be overemphasized. Once again, repulsion was associated with the absence of coeducation. The college boardinghouse was an accumulation of the mephitism of the walls, the social stench of the domestic staff, and the odor of the sperm of the schoolmaster and his masturbating pupils. This stench, perceived as male, sharpened desire for the presence of females.

32. Charles Baudelaire, *L'Invitation au voyage* (prose poem).

33. Jean-Pierre Richard, *Proust et le monde sensible* (Paris, 1974), 101.

34. Bachelard, *La Poétique de l'espace* (Paris, 1957), 32, 83, praised "the single cupboard, the cupboard with a single odor, which signifies intimacy," the cupboard, "center of order," with vegetable scents. With lavender, "the history of the seasons [entered] the cupboard. By itself lavender introduces a Bergsonian temporality into the hierarchy of the sheets. Before using them, does one not have to wait until they are, as we say, lavendered enough?" Given the association established between order and vegetable scent, the rejection of animal perfumes amounted primarily to a rejection of disorder.

35. See further discussion in Chapter 11, pp. 186–188.

36. Bachelard (*La Poétique de l'espace,* 44, 47, and 130) expanded on the theme of

the fundamental quality of the home, which intensified the "value of a center of concentrated solitude," stimulated the quest for "centers of simplicity" inside the house, and turned the smallest recess where the child could curl up into "the seed of a chamber." See also Chapter 12, pp. 207–208.

37. Howard, *Principal Lazarettos*, 20–21.
38. Londe, *Nouveaux Eléments*, 1:406, 407.
39. Odier, *Principes d'hygiène de Sinclair*, 577.
40. Guy Thuillier (*Pour une histoire du quotidien*, 41) has noted that until 1900 sweeping without water remained the rule in Nivernais schools.
41. Piorry (*Des habitations*, 34) furnished a list of these works, including several by Benoiston de Châteauneuf.
42. Forget, *Médecine navale*, 198. He added: "The broom will go into all the recesses, behind, between, and underneath the chests, which have to be moved for this purpose. It is the darkest and most concealed places that require the most supervision."
43. Howard, *Principal Lazarettos*, 202.
44. Tenon, *Mémoires sur les hôpitaux*, 186 ff.
45. See Denis I. Duveen and Herbert S. Klickstein, "Antoine Laurent Lavoisier's Contributions to Medicine and Public Health," *Bulletin of the History of Medicine*, 29 (1955), 169.
46. Béguin, "Evolution de quelques stratégies," 236.
47. Péclet, *Instruction*, 2; Leblanc, *Recherches*, 21.
48. Notably Passot, *Des logements insalubres*, 16.
49. Monfalcon and Polinière, *Traité de la salubrité*, 65.
50. Piorry, *Des habitations*, 89.
51. In the second half of the century, architects were less concerned with health and more with the pleasure of living. Hygiene was then considered only an ingredient of comfort. See A. Thalamy in Comité, *Politiques de l'habitat*, 50.
52. Quoted in ibid., p. 34.
53. Mille, "Rapport sur la mode d'assainissement," 224; François Béguin, "Les Machineries anglaises du confort," *L'Haleine des Faubourgs, Recherches*, 29 (1977), 155–186.
54. Mille, "Rapport sur la mode d'assainissement," 219, 221.
55. "I remember," noted Mrs. Trollope in 1835, "being much amused last year, when landing at Calais, at the answer made by an old traveller to a novice who was making his first voyage. 'What a dreadful smell!' said the uninitiated stranger, enveloping his nose in his pocket-handkerchief. 'It is the smell of the continent, sir,' replied the man of experience. And so it was"; *Paris and the Parisians in 1835*, 1:230.
56. Murard and Zylberman, "Hygiène corporelle et espace domestique, la salle de bains," in "Sanitas sanitatum," 292.
57. Piorry, *Des habitations*, 130, 131.
58. Grassi, *Rapport sur la construction*, 28.
59. Ibid., pp. 29, 30.
60. Numerous documents concerning this offensive by the Parisian administration can be found in Departément de la Seine, Ville de Paris, Commission des Logements Insalubres, *Rapport général des travaux . . . années 1862–1865* (1866).

A systematic campaign was waged against Turkish-style bowls (holes in the ground) and temporary latrines; town councillors pinned their hopes on schools. Norms were defined (p. 79). The administration's program provided for these latrines to be "installed in the uncovered yard, isolated, facing north, to the number of two per hundred pupils, suitably aired and ventilated," disinfected, and placed under the supervision of the caretaker, promoted general-in-chief in the war against excrement. The model remained the school at 77 rue de Reuilly because an old woman constantly cleaned the latrines there (p. 32). What is striking in this literature is the extraordinary detail of the proposals (p. 34). Progress was greater in secondary than in primary schools and more rapid in girls' than in boys' schools.

61. Ibid., p. 34.

62. Ibid., p. 29. See also Laporte, "Contribution," 224 ff., quoting a good but much later text on this subject. The inspectors' reports frequently allude to the unpleasant odor of schools; often they were sufficient to justify a decision to close the schools down.

63. Guerrand, "Petite Histoire du quotidien," 96–99.

64. Charles de Gaulle in his description of national temperaments notes the German habit of erecting "Gothic palaces for the needs of nature"; *Vers l'armée de métier* (Paris, 1971), 27 (1st ed. 1934).

65. Grassi, *Rapport sur la construction*, 29.

66. Lecadre, "Le Havre," 256–257.

67. A bathroom was installed on the first floor of the typical Lille house described by A. de Foville in 1894, at the time of the inquiry into living conditions; A. Thalamy in Comité, *Politiques de l'habitat*, 33.

68. Chaline, *La Bourgeoisie rouennaise*, 807.

69. As early as 1827, Antoine Caillot (*Mémoires*, 2:100) stressed the role of kept women in spreading the demand for bathrooms.

70. Alfred Picard, in République française, Ministère du Commerce et de l'Industrie, des Postes et Télégraphes, *Exposition universelle internationale de 1900 à Paris. Rapport général administratif et technique par M. Alfred Picard*, 8 vols. (Paris, 1901–3), 6:3.

71. Lawrence Wright, *Clean and Decent: The Fascinating History of the Bathroom and the Water Closets* (London, 1960). This work contains illustrations of the luxurious water closets of the Victorian period (p. 206). Acanthus leaves compete with blue magnolias in decorating the ceramics. The masterpiece appears to be a bowl whose pedestal consists of a sculptured lion.

72. See Murard and Zylberman, "Sanitas sanitatum," 291.

11. *The Perfumes of Intimacy*

1. The title of this section is taken from the comtesse de Bradi (née Agathe Caylac de Caylan; she was the pupil of Madame de Genlis), *Du savoir-vivre en France au XIXᵉ siècle* (1838), 210.

2. Duveen and Klickstein, "Lavoisier's Contributions."

3. A term introduced by Broussais.

4. The hygiene of the senses occupied an important place in hygiene manuals.

For example, Rostan (*Cours élémentaire d'hygiène*, 1:530) stressed the importance of the hygiene of the sense of touch.

5. "The complexion must always be a mixture of roses and lilies . . . if a pure color circulates beneath a white, fine, sweet, and fresh skin," decreed Louis Claye, *Les Talismans de la beauté* (1860), 90–91.

6. Ibid., p. 94. Jean-Pierre Richard, *L'Univers imaginaire de Mallarmé* (Paris, 1961), 92 and 61, has written fascinating articles about the "splendor of the original white" and the genesis in Eden of the white flower, which is linked with the eternal snow of the stars. The extent to which symbolism later contributed to relaunching this taste for pearly skin is well known. Mallarmé himself extolled the virtues of snow-cream.

7. Werner Sombart, *Le Bourgeois* (Paris, 1926), 134.

8. Madame Celnart, *Manuel des dames ou l'art de l'élégance* (1833), 100.

9. Vidalin, *Traité d'hygiène domestique*, 159.

10. Geneviève Heller, *Propre en ordre* (Paris, 1980), analyzed perfectly the convergent tactics that from 1850 on in the canton of Vaud tended to make Switzerland the land of cleanliness—the supreme virtue that hallowed all the others, since it implied perseverance. Heller proved that until World War I, efforts were concentrated more on domestic cleanliness than on cleanliness of the body. See also Marie-Hélène Guillon, "L'Apprentissage de la propreté corporelle à Paris dans la deuxième moitié du XIXᵉ siècle" (Mémoire of Diplôme d'Etudes Approfondies, Paris VII, 1981).

11. Richard Sennett on the "green disease," *Les Tyrannies de l'intimité* (Paris, 1979), 145.

12. De Bradi, *Du savoir-vivre*, 180.

13. D.-M. Friedlander, *De l'éducation physique de l'homme* (1815), 54.

14. In 1804 P.-J. Marie de Saint-Ursin advised: "When, hesitating between the delights of voluptuousness and the honor of virtue, the young girl, her complexion pale, lips colorless, eyes wet with involuntary tears, seeks solitude and takes pleasure in melancholy reveries; let a long hot bath soothe the causes of this erotic orgasm; let it demolish the forces of this privileged child of nature"; *L'Ami des femmes*, 169. Witness the transference from the "hygiene of coquetry" to the "hygiene of temperament."

15. A. Delacoux, *Hygiène des femmes* (1829), 223, 224.

16. Ibid., p. 226. Parent-Duchâtelet attributed prostitutes' stoutness to excessive bathing.

17. Rostan, *Cours élémentaire d'hygiène*, 1:507.

18. Celnart, *Manuel des dames*, 37.

19. Marie de Saint-Ursin, *L'Ami des femmes*, 117.

20. See Marie-Françoise Guermont, "La Grande Fille. L'image de la jeune fille dans les manuels d'hygiène de la fin du XIXᵉ siècle et du début du XXᵉ siècle" (Master's thesis, Tours, 1981).

21. De Bradi, *Du savoir-vivre*, 210.

22. See P. Perrot, *Les Dessus*, 228.

23. De Bradi, *Du savoir-vivre*, 191.

24. Celnart, *Manuel des dames*, 8–12. She also mentioned the practice of smearing the hair with egg yolk to remove grease. Dr. J.-P. Thouvenin recommended

washing the hair from time to time in warm soapy water; *Hygiène populaire à l'usage des ouvriers des manufactures de Lille et du département du Nord* (1842), 27.

25. Londe, *Nouveaux Eléments,* 2:5.

26. Celnart, *Manuel des dames,* 23.

27. From our viewpoint, the rapid adoption of this "invisible garment" (P. Perrot, *Les Dessus,* 259) was an event of the greatest importance.

28. See Thuillier, *Pour une histoire du quotidien,* 124 ff.

29. It seems that this was the case at Minot; see Verdier, *Façons de dire,* 111–112. For the adolescent girl, the pleasant smell emitted by new fabrics was one of the attractions of the winter apprenticeship at the couturier's (p. 215).

30. According to Thuillier (*Pour une histoire du quotidien,* 52) it was in general use among the Nevers bourgeoisie in 1900, as were sanitary towels. Bidets were used by other groups only after 1920.

31. Martin-Fugier, *La Place des bonnes,* 110.

32. Léonard, *Les Médecins,* 1468.

33. The advance was encouraged by the spread of enameled iron, which permitted cheap production of bowls of vast dimensions. The new requirements therefore created a generation gap.

34. See Thuillier, *Pour une histoire du quotidien,* 54–55.

35. Villermé, *Des prisons,* 34.

36. Fanny Faÿ-Sallois, *Les Nourrices à Paris au XIXe siècle* (Paris, 1980), 216.

37. Verdier, *Façons de dire,* 122–128. Thuillier has noted an analogous process in the Nivernais. A veritable "washhouse policy" was launched in the decade 1820–1830. Control of water in the rural communes advanced rapidly between 1840 and 1870. However, a coherent and systematic sanitary policy was not defined until the law of February 15, 1902. Thuillier, *Pour une histoire du quotidien,* 14 ff.

38. According to Madame de Girardin (*Lettres parisiennes,* 317), in Paris in 1837 even the fashionable gentleman exhaled a strong odor of tobacco. The tactful Paz, hero of Balzac's *La Fausse Maîtresse,* was afraid of infecting the comtesse Laginska's barouche with a bad smell, because he had just smoked a cigar (Paris, 1976), 2:218.

39. Veblen, *Theory of the Leisure Class,* 83, 179–180. See P. Perrot, *Les Dessus.*

40. In 1825 Madame Gacon-Dufour (*Manuel du parfumeur,* 31, 83) stressed the decline of musk and the primacy of eau de cologne and melissa cordial. Strong odors such as musk, ambergris, orange blossom, and tuberose should be completely forbidden, according to Madame Celnart in 1833 (*Manuel des dames,* 11).

41. Tourtelle, *Eléments d'hygiène,* 434.

42. De Bradi, *Du savoir-vivre,* 214.

43. Tourtelle, *Eléments d'hygiène,* 434–435. Rostan voiced the same opinion, *Cours élémentaire d'hygiène,* 1:528–529.

44. Eugène Rimmel, *Le Livre des parfums* (Brussels, 1870), 25.

45. Ibid., p. 350.

46. Claye, *Les Talismans de la beauté,* 75.

47. Louise de Chaulieu still used it to hold Marie Gaston in *Mémoires de deux jeunes mariées* (Paris, 1979), 381.
48. A. Debay, *Les Parfums et les fleurs* (1846), 49.
49. See Londe, *Nouveaux Eléments*, 2:501.
50. Madame Celnart (*Manuel des dames*, 92) indulgently also permitted "a few drops of eau de cologne" on bodice and stockings.
51. Londe, *Nouveaux Eléments*, 1:59.
52. De Bradi, *Du savoir-vivre*, 220. The same principles inspired the list of perfumes permitted by Delacoux in 1829 (*Hygiène des femmes*, 233) and by Madame Celnart in 1833 (*Manuel des dames*, 92).
53. Rimmel, *Le Livre des parfums*, 369.
54. Debay, *Les Parfums et les fleurs*, 42.
55. Rostan, "Odeur," in *Dictionnaire de médecine* (Béchet). See also Friedlander, *De l'éducation physique*, 70.
56. Dr. Z.-A. Obry, *Questions sur diverses branches des sciences médicales* (1840), 13. Madame Celnart translated the medical precepts for the benefit of her elegant readers: "Pallor, thinness, rings round the eyes, low spirits, and nervous shivers are the normal fruits of the exaggerated use of scents by people whose nerves are more or less irritable" (*Manuel des dames*, 91). A. Debay advised against the use of perfumed gloves, by themselves capable of causing accidents; *Hygiène des mains et des pieds, de la poitrine et de la taille* (1851), 20.
57. Dr. Alexander Layet, "Odeurs," in *Dictionnaire Dechambre* (1880).
58. See Antoine Combe, *Influence des parfums et des odeurs sur les névropathes et les hystériques* (Paris, 1905).
59. Mauzi, *L'Idée du bonheur*, 271.
60. Madame Celnart emphasized that expensiveness and discretion were inseparable where perfume was concerned. Vegetable odor dissipated more rapidly than animal scents. Accordingly, light perfume involved greater expense; its use was evidence of wealth.
61. Rostan, *Cours élémentaire d'hygiène*, 1:528.
62. See Borie, *Mythologies de l'hérédité*, 57.
63. See Michel Foucault, *La Volonté de savoir* (Paris, 1977).
64. Debay, *Les Parfums et les fleurs*, 50.
65. M. Barruel, "Mémoire sur l'existence d'un principe propre à caractériser le sang de l'homme et celui des diverses espèces d'animaux," *Annales d'Hygiène Publique et de Médecine Légale*, 1 (1829), 267–277.
66. Rostan, "Odorat," in *Dictionnaire de médecine* (Béchet) (1840).
67. Londe, *Nouveaux Eléments*, 1:59.
68. Cloquet, "Odeur," in *Dictionnaire des sciences médicales* (Panckoucke) (1819), 229.
69. Rostan, "Odorat," 237.
70. A perfume-pan was included in Louise de Chaulieu's trousseau (*Mémoires de deux jeunes mariées*, 213).
71. Chaptal, *Eléments de chimie*, 109.
72. Debay, *Les Parfums et les fleurs*, 43.
73. It is known that he drew his inspiration from Laure d'Abrantès's apartments. According to Antoine Caillot (*Mémoires*, 2:134) the Directory restored to the

boudoir all its previous importance, and in particular its political role. It was then that hairdressers became fashionable. "The whole woman is there . . . and in her bedroom," declared Baron Mortemart de Boisse in 1857 à propos the boudoir (*La Vie élégante à Paris,* 89).

74. De Bradi, *Du savoir-vivre,* 221. "They greeted them with rapture like so many sisters happily refound"; Jules Janin, *Un Eté à Paris* (1844), 238.

75. Quoted in Raymond, *Senancour, sensations et révélations,* 157.

76. Michelet, *La Femme,* 242–243.

77. Ingenhousz, *Expériences sur les végétaux,* lxxxviii.

78. Michelet, *La Femme,* 127, 128.

79. At the most it is possible to discern an outline of an evolution. Theorists ceased to rely on nature and advised strewing lawns with odoriferous flowers: irises, lilies of the valley, violets, meadow geraniums. More specific attention was paid to aromas because of the stress on everything related to respiration. "On the banks of the river grow aromatic plants, wholesome herbs; their balsamic odor, joined with the odor of the resin from the odoriferous pines, perfumes the air and dilates the lungs"; J. Lalos, *De la composition des parcs et jardins pittoresques* (1817), 88.

80. On their popularity in England, see Edmond Texier, *Tableau de Paris,* 2 vols. (1852–53), 1:154.

81. Comte Alexandre de Laborde, *Description des nouveaux jardins de la France et de ses anciens châteaux* (1808), 210.

82. In 1858 Mortemart de Boisse ended his description of the elegant woman's apartment in this way: "All the casements on the ground floor give onto a greenhouse-garden which, four or five times during the winter, tapestrymakers transform into a small theater where society men and women play at proverbs"; *La Vie élégante,* 90.

83. Charles-François Bailly, *Manuel complet théorique et pratique du jardinier,* 2 vols. (1829), 1:223.

84. Baron Alfred-Auguste Ernouf, *L'Art des jardins,* 3d ed. (1886), 238.

85. See Edouard André, *Traité général de la composition des parcs et jardins* (1879), 192.

86. Bory de Saint Vincent, *Musée des familles,* vol. 1 (1834), quoted in Arthur Mangin, *Histoire des jardins, anciens et modernes* (1887), 372. The precocity of Bory de Saint Vincent's description is noteworthy. It remains very far from Zola's model of the woman as a poisonous liana and farther still from the symbolistic scenes of the end of the century. The greenhouse that Sombreval installed at Quesnay for the benefit of his "sensitive" Callixte respected the discretion in force at the beginning of the century; Barbey d'Aurevilly, *Un Prêtre marié* (1865).

87. Pierre Boîtard, *L'Art de composer et décorer les jardins,* 2 vols. (1846), 2:22.

88. J.-C. Loudon, *Traité de la composition et de l'exécution des jardins d'ornement* (130), 194. Madame Jean-Marie Roland de la Platière reported such a separation in the garden of her childhood; *Mémoirs particuliers* (Paris, 1966), 205 (1st ed. 1847).

89. Laborde, *Description des nouveaux jardins,* 210.

90. Bailly, *Manuel complet théorique,* 2:47.

91. See, for example, the role of the garden in the life of Duranty's heroine in *Le Malheur d'Henriette Gérard*. The description of the young girl's awakening is revealing. "She got up, heard the birds singing, smelled the flowers, looked at the changing sky"; (Paris, 1981) 112.
92. Bailly, *Manuel complet théorique*, 2:57.
93. The following definitions are from Boîtard, *L'Art de composer*.
94. Note the birdsong in Modeste Mignon's garden.
95. Michelet, *La Femme*, 129.
96. Madame Marie Fortunée Lafarge, *Heures de prison* (1853), 92. The sweet odor of reseda flowers, which she had inhaled and chewed while playing, enabled Madame de Stasseville to visualize a child's corpse buried in the open ground in the box where the plant grew. This odor was so intrusive in her salon that delicate ladies refused to frequent it; Barbey d'Aurevilly, *Les Diaboliques* (Paris, 1973), 219.
97. "Who has not got a profusion of violets in his garden?" asked Bailly (*Manuel complet théorique*, 2:174); julienne was also "one of the plants most used for the decoration of flowerbeds." The great success of what was called "ladies' stock" or, even more, "garden rocket" was also due to its magic perfume. On the other hand, tuberoses were distrusted.
98. See Pierre Boîtard, *Le Jardinier des fenêtres, des appartements et des petits jardins* (1823).
99. De Bradi, *Du savoir-vivre*, 221.
100. Marcel Détienne has described the sham cultivation practiced by Greek women in the gardens of Adonis installed on their terraces like an illusory agriculture, the antithesis of the cultivation of grains. In the nineteenth century, the pastime of growing things in flowerbeds and pots, popular among women of the elite, could symbolize the uselessness of the female's time, fortunately counterbalanced by her husband's productive activity.
101. Madame Amet née Abrantès, *Le Messager des Modes et de l'Industrie*, March 1, 1855.
102. "A charming coiffure worn by the empress quite recently," wrote Madame Amet, "was a plait of hair placed over the forehead and braided with natural flowers. They were buds of large white daisies"; ibid.
103. Mrs. Trollope, *Paris and Parisians in 1835*, 2:6. In his *Tableau de Paris*, published in 1852, Texier emphasized at length and precisely the expansion of the flower trade as well as the splendors of the winter garden. Evenings in the Mabile garden seemed to him more fragrant than in the past. "The harmony of Pilodo's orchestra blended voluptuously with the scent of jasmine and roses," noted Madame Amet (*Le Messager des Modes et de l'Industrie*, July 15, 1855). Wherever it moved, the imperial feast unfolded in a debauchery of sweet perfumes.
104. See Davin, "Le Printemps à Paris," in *Le Nouveau Tableau de Paris*, vol. 1 (1834), 209.
105. Debay, *Les Parfums et les fleurs*, 216.
106. Paul de Kock, "Les Grisettes," in Davin, *Le Nouveau Tableau de Paris*, 174. Davin stated that sweet peas and particularly reseda were "cherished" by the grisette and by the housewife, who "voluptuously perfumed her stomach with

them." The moment she got up, the young girl ran to her little garden ("Le Printemps à Paris," 211). In 1852 Texier mocked the grisette's taste for reseda, the student's for violets; "the sentimental foot soldier" preferred to offer his countrywoman a pot of cloves; *Tableau de Paris*, 1:153.

107. See Zylberberg-Hocquard, "L'Ouvrière dans les romans populaires," 614, on the importance given to flowers and birds by authors of popular novels.

108. See Verdier, *Façons de dire*, 185.

109. Cf. also Serge's pastoral at the beginning of *La Faute de l'abbé Mouret*.

110. Victor Hugo, *Toilers of the Sea*, 3 vols. (London, 1866), 1:136.

111. Ibid., 3:189–190.

112. Ibid., 1:180–181.

113. Honoré de Balzac, *The Country Doctor* (London, 1911), 122.

114. André, *Traité général*, iii.

115. Ibid., pp. 687–717.

116. A. Alphand and A.-A. Ernouf, *L'Art des jardins* (1886), 326.

117. In the countryside (see the cycle of *Claudine* by Colette), the innocent alliance between the young girl and the flower persisted, in contrast to the development of Parisian fashions. Symbolist art also continued to refine the parallelism between the young girl and the sweet flower. Theodor Fontane's Romantic work, notably the subtle floral symbolism of Effi Briest's garden, is revealing on this subject.

118. Claye, *Les Talismans de la beauté*, 24.

119. Claude Rifaterre suggested that the term *muscadin* initially (August 1792) designated grenadiers of the Lyons national guard, young men of good social position and shop and bank clerks, looked on unfavorably by the sans-culottes, who formed the military groupings in the center of the city. The term was immediately and proudly revived by interested parties; "L'Origine du mot muscadin," *La Révolution Française*, 56 (January–June 1909), 385–390.

120. Madame Celnart, *Manuel de parfumeur* (1834), 225.

121. Claye, *Les Talismans de la beauté*, 35.

122. Pfeiffer, *Taste and Smell*, 27.

123. Dumas, "Les Parfums," *Le Moniteur Universel du Soir*, October 12, 1868.

124. De Bradi, *Du savoir-vivre*, 211.

125. Chaulieu, *Mémoires*, 1:200.

126. A. Debay, *Nouveau Manuel de parfumeur-chimiste* (1856), 40.

127. Madame de Girardin (*Lettres parisiennes*, 329) dated the abandonment of stiffness, the challenge to elegant simplicity, and the return to fantasy very precisely to 1839. Despite the demise of horticulture, the author remained loyal to the sweet odors of jasmine and honeysuckle.

128. See Vigarello, *Le Corps redressé*, 167.

129. The rejection of ambergris and musk remained *de rigueur* at the imperial court, evidence of good taste and morality. The composition of the Bouquet de l'Impératrice that Guerlain prepared for the sovereign is revealing on this subject. Although the perfume used by Queen Victoria on her official visit to France in 1855 was of high quality, it still included a discrediting trace of musk. The fashionable ladies of the Tuileries eagerly drew attention to it; Madame Amet née Abrantès, *Le Messager des Modes et de l'Industrie*, June 1, 1855.

130. These findings are from a quantitative study the details of which have no bearing here.

131. See M.-L. L'Hôte, *Rapports du jury international, publiés sous la direction de M. Alfred Picard* (Paris, 1891).

132. Claye, *Les Talismans de la beauté*, 56.

133. See Albert Boime, "Les Hommes d'affaires et les arts en France au XIXe siècle," *Actes de la Recherche en Sciences Sociales*, 28 (June 1979).

134. Rimmel, *Le Livre des parfums*, 24.

135. P. Perrot, *Les Dessus*, 325–328.

136. S. Piesse, *Des odeurs, des parfums et des cosmétiques*, 2d ed. (1877), 4–18 (1st ed. London, 1855, *The Art of Perfumery*).

137. See Debay, *Nouveau Manuel du parfumeur-chimiste*, 107.

138. For their bandolines and lustrines alone, Gellé frères in 1858 suggested flat, square, round flasks, "tomb," "violin," "stag-beetle," "in a case," "gourds"; survey in the series *Parfumeries* (Bibliothèque Nationale, V. 403), a collection of prospectuses of different firms in the nineteenth and twentieth centuries.

139. On the concept of the shape of smells, phraseology of smells, and the perfume-composer, see the very fine book by O. Moreno, R. Bourdon, and E. Roudnitska, *L'Intimité des parfums* (Paris, 1974).

140. Including Lane's *Modern Egyptians* (1837), Sonnini's *Voyage en Egypte* (1799), and Duckett's *La Turquie pittoresque* (1855). The reconstruction of the Palace of the Bardo at the Exposition of 1867 would also have contributed to the fashion for the Orient, already relaunched by the Crimean War. It was to this latter episode that Madame Amet attributed the still discreet revival of the use of makeup.

141. Flaubert, *Correspondance*, 1:558 (January 5, 1850) and 568 (January 15, 1850) (Paris, 1973).

142. Goncourt and Goncourt, *Manette Salomon*, 131.

143. Léonard, *Les Médecins*, 1468. The flasks of perfume that decorated the office of the mayor of Plassans impressed Antoine Macquart; they made him aware of the social distance between himself and Rougon and finally calmed the violence of his revolt; Emile Zola, *La Fortune des Rougons* (Paris, 1960), 271–272.

12. *The Intoxicating Flask*

1. See Charles de Rémusat, *Mémoires de ma vie*, vol. 1 (Paris, 1958), 110 ff.

2. Balzac, *Les Paysans*, 53.

3. Balzac, *The Country Parson* (London, 1914), 17–18.

4. See the account of his sister's wake in his *Selected Letters*. At the same time, the elegant ladies of Corrèze came in a crowd to breathe the stench that emanated from poor Lafarge's guts in the Tulle court.

5. "Nature embracing all his senses . . . he forgot himself, lost himself, in seeing, in listening, in aspiring . . . There is the scent of virginias in flower in the air that Anatole breathes . . . There are steaming aromas, musky emanations, and wild odors mingled with the sweet scents from the bushes of *cuisse de nymphe* roses that perfume the entrance to the garden"; *Manette Salomon*, 425.

6. Maine de Biran, *Journal*, 79.

7. Ibid., pp. 77, 165.

8. Senancour, "Promenade en Octobre," *Le Mercure du XIXᵉ siècle,* 3 (1823), 164.

9. Maine de Biran, *Journal,* 152.

10. Cloquet, "Odeur," 229.

11. Cloquet, *Osphrésiologie,* 112.

12. Dr. Bérard, "Olfaction," in *Dictionnaire de médecine* (Béchet), 19.

13. Balzac, *Louis Lambert,* p. 607.

14. George Sand, *Histoire de ma vie,* vol. 1 (Paris, 1970), 557.

15. Charles Baudelaire, "Le Parfum."

16. Alphonse Karr, Preface to Eugène Rimmel, *Le Livre des parfums* (Paris, E. Dentu, n.d.).

17. G. Flaubert, *Madame Bovary* (Paris: Editions Pléiade, 1951), 473. Captain Bertin, the hero of *The Master Passion,* provides another example of the complex memory. At sea, when he breathed the odor of his native Corsica, he was assailed by "those vanished memories, which, seemingly lost for ever, have a trick of suddenly returning, no one knows why. They came swiftly crowding upon him, so had stirred up the dregs of his subconscious mind. He tried to discover the reason of this fermentation of his old life . . . There was always something to account for these sudden evocations, some simple natural cause, more often than not a scent, a fragrant odour. How often the flutter of a woman's gown, wafting to him, as she passed by, the airy effluence of some perfume, had conjured up a whole sequence of forgotten incident. From old empty scent bottles he had often recovered some reminiscence of the past, while all the stray odours, pleasant and unpleasant, of streets, fields, houses, furniture, the warm scent of summer evenings, the chill breath of winter nights, always revived for him memories of long ago"; Guy de Maupassant, *The Master Passion,* trans. Marjorie Laurie (London, 1961), 54.

18. Fromentin, *Dominique,* trans. V. I. Longman (London, 1932), 85.

19. This potential was understood by perfumers, who offered flasks shaped like tombs enclosing the perfume of the vanished woman.

20. Eugène-Emmanuel Viollet-le-Duc, *Dictionnaire raisonné de l'architecture français du XIᵉ au XVIᵉ siècle,* vol. 6 (1859), 164. Frédéric and Rosanette walking at Fontainebleau saw these "exhalations of the centuries"; see Richard, *Littérature et sensation,* 190.

21. Théophile Gautier, "Le Pied de momie" and "Arria Marcella," in *Récits fantastiques* (Paris, 1981), 184, 251.

22. Charles Baudelaire, "The Perfume Flask," in *The Flowers of Evil,* trans. George Dillon and Edna St. Vincent Millay (London, 1936), 9.

23. Zola, *La Joie de vivre,* 857.

24. T. Thoré, *Dictionnaire de phrénologie et de physiognomonie à l'usage des artistes, des gens du monde, des instituteurs, des pères de famille, etc.* (1836), 314.

25. Barbey d'Aurevilly was devastated by the idea of the odor of the beloved woman struggling against illness, as if these effluvia bore witness to distress more than anything else. "It is necessary to have smelled around her poor fevered face the breaths laden with life from the gown that encloses the woman one loves"; *Un Prêtre marié,* 233.

26. Sainte-Beuve's novel *Volupté* seems to have established a model in this respect that inspired George Sand *(Lélia)* and Balzac *(Le Lys dans la vallée).*

27. Pfeiffer, *Taste and Smell*, 49.
28. Balzac, *Le Lys dans la vallée*, 1114.
29. Baudelaire, "The Fleece," in *The Flowers of Evil*, 23.
30. Baudelaire, "Chanson d'après-midi."
31. See Baudelaire, "La Propreté des demoiselles belges."
32. Precisely described by Maupassant, *L'Ami Patience*.
33. Baudelaire's infinite variations on perfume, imaginary journeys, correspondences, and reminiscences are outside my subject. But his quest for supreme ecstasy when all the senses were overwhelmed can be regarded as the culmination of a long process, traced by, among others, M. A. Chaix, *La Correspondance des arts dans la poésie contemporaine* (Paris, 1919), and Jean Pommier, *La Mystique de Baudelaire* (Paris, 1932). The theme of the eternality of perfume also obsessed Baudelaire's contemporaries; and the incitement that smells offered to travel belonged to the collective imagination (see Chapter 11, p. 199).
34. The role of olfaction in Zola's works has already been studied by others.
35. Dr. Edouard Toulouse, *Enquête médico-psychologique sur les rapports de la supériorité intellectuelle avec la névropathie. Emile Zola* (1896), 163–165, 173–175.
36. Léopold Bernard, *Les Odeurs dans les romans de Zola* (Paris, n.d.).
37. Alain Denizet, *Les Messages du corps dans les Rougon-Macquart* (Master's thesis, Tours, 1981).
38. Bernard, *Les Odeurs dans Zola*, 8.
39. Typical in this respect was the men's conversation in stage whispers at the comtesse Muffat's soirée (*Nana*, chap. 3).
40. See, for example, Dr. Deberle's seduction of Hélène Grandjean in Zola's *Une Page d'amour*.
41. See Richard, *Littérature et sensation*, 189.
42. Flaubert, *Selected Letters*, 77 (Aug. 8), 80 (Aug. 9), 88 (Aug. 15); *Letters of Flaubert to Louise Colet*, trans. F. Steegmuller (London, 1980), 58 (Aug. 11), 62 (Aug. 13).
43. Zola, *Zest for Life*, trans. J. Stewart (London, 1955), 196–197.
44. Summarized by Havelock Ellis, *Sexual Selection in Man*, 169 ff. Hagen thought that the odor of leather recalled the odor of the sexual organs.
45. Edmond Huot de Goncourt, *Chérie* (1889).
46. Auguste Ambroise Tardieu, *Les Attentats aux moeurs* (1867), 183.
47. G. Macé, *La Police parisienne. Un joli monde* (1887), 263, 266, 272.
48. Charles Féré, *La Pathologie des émotions* (1892), 438–441, and *L'Instinct sexuel. Evolution et dissolution* (1890), 126 ff. and 210 ff.
49. Alfred Binet, "Le Fétichisme dans l'amour," in *Etudes de psychologie expérimentale* (1888), 4. He recalled that for Morel, as for Magnan, these defects were only episodes in the hereditary madness of degenerates. What seems to have been essential for Binet was the fact that in the olfactive fetishist, odor unleashed an irresistible impulse; it made him follow the woman whose effluvia fascinated him. According to Féré (*La pathologie des émotions*, 439), Lamartine probably liked girls from inns for this reason.
50. Joris-Karl Huysmans, *Against the Grain*, trans. R. Baldick (Paris, 1926), 163 (1st ed. 1884).
51. See Pierre Cogny, "La Destruction du couple Nature-Société dans l'*A rebours*

de J.-K. Huysmans," and Françoise Gaillard, "De l'antiphysis à la pseudo-physis: L'exemple d'*A rebours*," *Romantisme*, 30 (1980).

52. Gaston Leroux, *Le Mystère de la chambre jaune* (Paris, 1960), for example, 84.

53. Jean Lorrain, *La Ville empoisonnée* (Paris, 1936), 106–107. On "the stenches of the black village": "And the odor of the Negro, an unpleasant smell of salty butter and pepper, rises, more nauseatingly on stormy nights." Clearly, the tone has changed.

54. Dr. Edgar Bérillon, "Psychologie de l'olfaction: La fascination olfactive chez les animaux et chez l'homme," *Revue de l'Hypnotisme*, October 1908, pp. 98 ff. This article expresses the idea—a source of considerable anxiety at the time—that civilization entails degeneration. It analyzes the decline in the role of olfaction in that perspective. But Bérillon was also very conscious that a return to a large-scale use of the sense of smell might signify regression; we thus see yet again the narrow dividing line.

55. Bérillon, "Psychologie de l'olfaction," 306. Dr. Bérillon rose to fame in 1915 with the publication of *La Bromidrose fétide de la race allemande, foetor germanica*, after having enlarged on the theme of *foetor judaicus*. The importance later given to racial odor in William Faulkner's *Intruder in the Dust* is well known.

13. *"Laughter in a Bead of Sweat"*

1. The title of this chapter is taken from J.-K. Huysmans's phrase; see p. 199.
2. See Corbin, *Archaïsme et modernité*, 337–362.
3. Chauvet, *Essai sur la propreté*, 7, 8. The Spanish government consulted the universities of Europe on the effects of this stench.
4. Fourcroy, *Essai de Ramazzini*, 561.
5. R.-P. Cotte, "Air et atmosphère," in *Encyclopédie méthodique. Médecine* (1787), 587.
6. For example, Alexandre Parent-Duchâtelet, "Essai sur les cloaques et égouts de la ville de Paris," in *Hygiène publique*, 1:252.
7. Parent-Duchâtelet, *Recherches pour découvrir la cause d'accidents*, 274.
8. Michel-Augustin Thouret, *Supplément au rapport sur la voirie . . .* (1788), 26.
9. Liger, *Fosses d'aisances*, 12.
10. Bailly, *Manuel complet théorique*, 2:586.
11. Parent-Duchâtelet, *Les Chantiers*, 139–140, n. 40.
12. Isidore Bricheteau, A. Chevallier, and Salvatore Furnari, "Note sur les vidangeurs," *Annales d'Hygiène Publique et de Médecine Légale*, 28 (1842), 50.
13. See Chapter 7, p. 119, on Chevreul; and Moléon, *Rapports généraux*, 2:495 (report for 1839).
14. Bertherand, *Mémoire*, 7, and Pierrard, *La Vie ouvrière*, 54.
15. "It is better to die of cholera than of hunger," a peasant told the mayor of the small commune of Saint-Priest-Ligoure in the Haute-Vienne; according to this magistrate, it was unfeasible to remove dungheaps; Corbin, *Archaïsme et modernité*, 77.
16. Alain Faure, *Paris Carême-prenant* (Paris, 1978), 107.
17. Gisquet, *Mémoires*, 458–465.

18. See Françoise Dolto, "Fragrance," *Sorcières*, 5 (n.d.), 12, and 10–17 for the discussion in this paragraph.
19. Verdier, *Façons de dire*, 329.
20. See Pierre Bourdieu, *La Distinction* (Paris, 1978), 574.
21. Faure, *Paris Carême-prenant*, 167. At Lille (Pierrard, *La Vie ouvrière*) the administration struggled for decades against "the pissers of the palisades" (p. 148). The first urinals installed in the city, during the Second Empire, inspired irony: using them was called "pissing in Paris fashion" (p. 53). In 1881 the Conseil d'Hygiène of the Seine could still report: "The floors of the lavatories are looked after, not the bowls . . . What is lacking . . . is the feeling, we would rather say the instinct, for cleanliness" (p. 284). But did this exist?
22. Faure, *Paris Carême-prenant*, 74.
23. Laporte, *Histoire de la merde*, 27.
24. Clocquet, *Osphrésiologie*, 115.
25. Fodéré, *Traité de médicine légale*, 6:539.
26. Ingenhousz, *Expériences sur les végétaux;* see above, pp. 36–37.
27. Howard, *Principal Lazarettos*, 215.
28. Cited in Bayard, *Mémoire sur la topographie médicale*, 88.
29. Quoted in François Béguin, "Savoirs de la ville et de la maison au début au XIXᵉ siècle," in Comité, *Politiques de l'habitat*, 259. Thuillier (*Pour une histoire du quotidien*, 39) emphasized how obstinately the workers of the Nivernais still demanded to work in very snug rooms at the beginning of the twentieth century. He stressed the need for a history of this resistance, which sealed the bond between employers and workers and which helps explain the failure of health policy.
30. Howard, *Principal Lazarettos*, 53.
31. Ibid., p. 161, and *State of Prisons*, 163.
32. Oliver Faure, "Hôpital, santé, société: Les hospices civils de Lyon dans la première moitié du XIXᵉ siècle," *Bulletin du Centre d'Histoire Economique et Sociale de la Région Lyonnaise*, 4 (1981), 45–51.
33. Sigmund Freud, *The Interpretation of Dreams*, ed. James Strachey (London, 1971), 238–239.
34. Bordeu, *Recherches*, 426.
35. Howard, *Principal Lazarettos*, 141; but the point concerns an opinion that was widespread among the masses in the following century; see, for example, Corbin, *Archaïsme et modernité*, 80.
36. Françoise Loux and Philippe Richard, *Sagesse du corps* (Paris, 1978).
37. What François Beguin ("Savoirs de la ville," 257) defines in these terms underlay a complex body of practices. Love of alcohol, tolerance of promiscuity, a preference for not working, ease of sexual contact, wandering the streets, and the search for anonymity all played a part. "Primitive bodily comfort" implied that dirt was better tolerated than effort. It supported confusion and strength of smells; it opposed the reforms of the "comfort" economy that authorities were trying to impose. "One walks comfortably only if one's thighs touch," Jules Renard's Ragotte declared, and thus justified her refusal to wear underpants; Thuillier, *Pour une historie du quotidien*, 58.
38. Boltanski, *Prime Education*, 83 ff.

39. Limousin proverb, in Corbin, *Archaïsme et modernité*, 81.
40. Françoise Loux has shown the utility of some of these prohibitions. For example, the refusal to remove dirt from children's heads corresponded to the wish to protect the fontanelle; *Le Jeune Enfant et son corps dans la médecine traditionnelle* (Paris, 1978).
41. Lapouge, "Utopie et hygiène," 104; Ariès, *L'Homme devant la mort*, 472.
42. Alain Corbin, "La Vie exemplaire du curé d'Ars," *L'Histoire*, 24 (May 1980), 7–15.
43. Lapouge, "Utopie et hygiène," 108.
44. Flaubert, *Correspondance*, 1:97.
45. Flaubert to Ernest Chevalier, October 23, 1841; *Correspondance*, 1:86.
46. See Jean-Paul Sartre, *L'Idiot de la famille* (Paris, 1971), 3:523.
47. See Lapouge, "Utopie et hygiène," 111.
48. Vallès, *L'Enfant*, 102.
49. Ibid., pp. 257, 321.
50. Stressed by Beatrice Didier, Introduction to *L'Enfant*.
51. Vallès, *L'Enfant*, 87.
52. Ibid., p. 73.
53. Ibid., pp. 87–88.
54. Unless Vallès exaggerated his rebellions as a child in order to show that the political revolt of his mature years was rooted in his childhood.
55. Vallès, *L'Enfant*, 89.
56. On the Republican printers in the rue Coq-Héron: "It is as good as the smell of dung. It smells as warm as in a stable"; ibid., p. 373.
57. Henry Miller, *Tropic of Capricorn* (London, 1971), 119–121; Günter Grass, *The Tin Drum*, passim. On the other hand, Bloom's association of ideas concerning the role of female odors (James Joyce, *Ulysses* [London, 1936], 356–357, 690–691) constitutes a long catalog of stereotypes; the Dublin petit bourgeois knew nothing of the "libertinage of the nose."

14. The Odors of Paris

1. Trélat, "Rapport," 25.
2. For example, Chrétien, *Les Odeurs de Paris*, 8.
3. Ibid., pp. 10 ff., and Alfred Durand-Claye, *Observations des ingénieurs du service municipal de Paris au sujet des projets de rapport présentés par MM. A. Girard et Brouardel* (1881).
4. With the one exception that, for a brief period, the microbe was sometimes called the microbian miasma.
5. Paul Brouardel, in *De l'évacuation des vidanges*, 36.
6. Dr. François-Franck, "Olfaction," in *Dictionnaire Dechambre* (1881), 99.
7. Marié-Davy, in *De l'évacuation des vidanges*, 65.
8. Quoted in Ariès, *L'Homme devant la mort*, 533.
9. Marié-Davy, in *De l'évacuation des vidanges*, 64.
10. Trélat, "Rapport," 19.
11. Durand-Claye, *Observations des ingénieurs*, 21–22.
12. Ibid., p. 23.

13. Ibid., p. 50.
14. Marié-Davy, in *De l'évacuation des vidanges,* 69.
15. Ibid.
16. See Murard and Zylberman, "Sanitas sanitatum."
17. Marié-Davy, in *De l'évacuation des vidanges,* 68.
18. See Alain Corbin, "L'Hérédosyphilis ou l'impossible rédemption," *Romantisme,* 1 (1981), 131–149.
19. O. Boudouard, *Recherches sur les odeurs de Paris* (Paris, 1912), 6. The author quoted an 1899 report by the Service for the Inspection of Classified Establishments.
20. There were eleven at Aubervilliers, two at Saint-Denis, three at Ivry, two at Vitry, and one in Paris; Paul Brouardel and Ernest Mosny, *Traité d'hygiène,* vol. 12 (1910), 161.

INDEX

Académie Royale des Sciences, 2, 24, 51, 59, 107

Acidic odor, 4, 19, 129

Adam (chief inspector of public buildings), 227

Aerotherapy, 78, 98, 215–217

Aesthetics, 61, 185, 198, 206, 229, 235n6; of flowers, 7, 190; of body, 169, 177

Age, and odor, 38, 140, 144, 188. *See also* Children, body odor; Puberty; Young girls, body odor

Agricultural workers, 23

Air, 11, 33, 113; physical qualities, 11–13, 14–15, 55, 62, 70; and foreign particles, 12, 13; quality, 39, 48, 52–53, 99, 114, 162; in valleys, 47, 78, 155. *See also* City air; Country air; "Fixed air"; Fresh air; Pure air; Respiration; Vapors

Air circulation, 91–92, 94, 96–97, 113, 123–124, 162, 165, 166. *See also* Ventilation

"Airs." *See* Gases

Almshouses, 216

Ambergris, 16, 69, 195, 196, 236n14; therapeutic effect, 62, 63, 67; discredited, 73, 74, 76, 182, 283n129

Amiens, 92

Ammonia, 112, 119

Ammonia salt, 120, 129, 130

Amsterdam, 217

Andral, Gabriel, 154

André, Edouard, 195

Anemometer, 125

Animal behavior, 6, 68, 111, 187; in humans, 6–7, 144, 153, 170, 186, 209

Animal black, 120, 123

Animal odors, 7, 44, 46, 49, 140, 150, 156, 187, 220; therapeutic effect, 37, 215

Animal perfumes, 67, 196, 205, 229; discredited, 49, 67, 68, 69, 73, 74, 147, 183, 184, 185. *See also* Ambergris; Civet; Musk

Animals, 110, 222–223; in dwellings, 161, 167, 215

Animal spirits, 62

Anthropology, 41, 140, 143, 144, 232; on race, 38, 39, 54, 187, 209–210

Antimephitic substances, 2–3, 91. *See also specific substances, e.g.,* Lime

Antiseptics, 17–18, 63, 102

Arbuthnot, John, 13, 95, 171

Architects, 98–99, 107, 126–127, 276n51

Architecture, 98–99, 107, 161, 171–172, 189. *See also* Buildings, construction; Landscape architecture

Aristocracy, 185, 189, 196, 198, 199, 214; sensitivity to odors, 203

Army, 101. *See also* Barracks; Soldiers

Aromatics, 67, 68, 112; therapeutic effect, 17, 18, 61, 62, 63, 65–66, 104; discredited, 69–70

Asphyxiation, 51, 60, 124, 166–167, 217; fear of, 112, 152–153, 162–164

Astringents, 18, 63, 66